A Bibliography of Arizona Ornithology

A Bibliography of *Arizona Ornithology*

ANNOTATED

Anders H. Anderson

THE UNIVERSITY OF ARIZONA PRESS
Tucson, Arizona

About the Author . . .

ANDERS H. ANDERSON has been published in many of the journals of ornithology, and is author, with his wife Anne, of the book *The Cactus Wren* (University of Arizona Press, 1972), based on a 30-year study of this state bird of Arizona. Born in Sweden, he came to the United States with his parents at the age of four, then later established deep-seated roots in the Southwest.

THE UNIVERSITY OF ARIZONA PRESS

Manufactured in the U.S.A.

I.S.B.N.-0-8165-0313-3
L.C. No. 76-163008

To Anne

Contents

Preface

This bibliography began insidiously many years ago as an attempt to gather into my library as many as possible of the published references to the birds of Arizona. Gradually, as the scope of this formidable task became evident, my efforts turned to the accumulation of entries on cards.

The last available list of publications relating to Arizona ornithology appeared in 1914. Harry S. Swarth in "A Distributional List of the Birds of Arizona" (Pacific Coast Avifauna 10: 1–133), included a compilation of about three hundred titles (pp. 94–119).

It has been customary in bibliographies to exclude the mimeographed materials, chiefly, I believe, because of their limited distribution, so-called ephemeral nature, and frequent unavailability. Most of the National Parks and National Monuments in Arizona have mimeographed lists of their local birds. These can be obtained at the various headquarters, but are seldom found in libraries. Reluctantly, I have excluded these, as well as the mimeographed publications of the local bird societies. For the sake of completeness, articles from journals such as *Grand Canyon Nature Notes* and *News from the Bird-Banders* (later the *Western Bird Bander*), originally mimeographed, but since changed to offset or other more desirable forms, are included.

I have excluded all newspaper articles and announcements of game laws. Popular materials, when they contain good photographs or colored illustrations of Arizona birds, are listed in the Semi-popular section. The section on Archaeology is probably weak; more titles, I feel sure, can be found in the voluminous literature of early man and excavations of prehistoric sites in Arizona. Papers on conservation and game management are considerably dispersed; some may have been overlooked. A few of the sources listed may contain only passing references to Arizona birds but are included intentionally for the benefit of others who share interests to the same degree as I do. For example, a paper on Mexican or Alaskan ornithology, offering minor, incidental references to distribution, measurements, or habits of Arizona species of birds, will be found here.

Beginning in 1948, the National Audubon Society published each year in *Audubon Field Notes* four season summaries for the "Southwest Region" of North America: Spring Migration, Nesting Season, Fall Migration, and Winter Season. The area covered includes all of Arizona, a narrow strip of California bordering the Colorado River, the southern tip of Nevada along with the southwestern corner of Utah, most of New Mexico, the western tip of Texas, and the northern portions of the Mexican states of Sonora and Chihuahua. The summaries, from contributions of numerous collaborators, have had remarkably few changes in authorship since the series began. In addition, beginning in 1966, from one to three comprehensive seasonal summaries of all of the North American regions have been included. Since all these summaries can be found readily in *Audubon Field Notes,* they have not been listed individually in this bibliography.

With extremely few exceptions, all references up to the end of August 1970 have been verified by a personal examination of the original publication or of reprints. When titles appeared to need amplification or explanation, I added brief annotations to the bibliographical entries.

Originally, I had decided on cross references between the main divisions of the bibliography, but I found so much overlap in the entries that the plan became undesirable. Almost all entries in the book can be considered to pertain to Section 2 in one way or another. Most of Game Management and Conservation is also Ecology.

I am grateful to numerous librarians from coast to coast, and particularly to those of the University of Arizona for frequent help. I thank Gale Monson, Allan R. Phillips, Stephen M. Russell, and Lyle K. Sowls for suggestions and help in locating publications, Enid K. Austin for the loan of bird banding journals, and Norman L. Ford for the use of important papers in the Josselyn Van Tyne Memorial Library. The encouragement and assistance of Anne Anderson made the work enjoyable and possible. I also thank the University of Arizona Press for bringing the bibliography into published form.

ANDERS H. ANDERSON

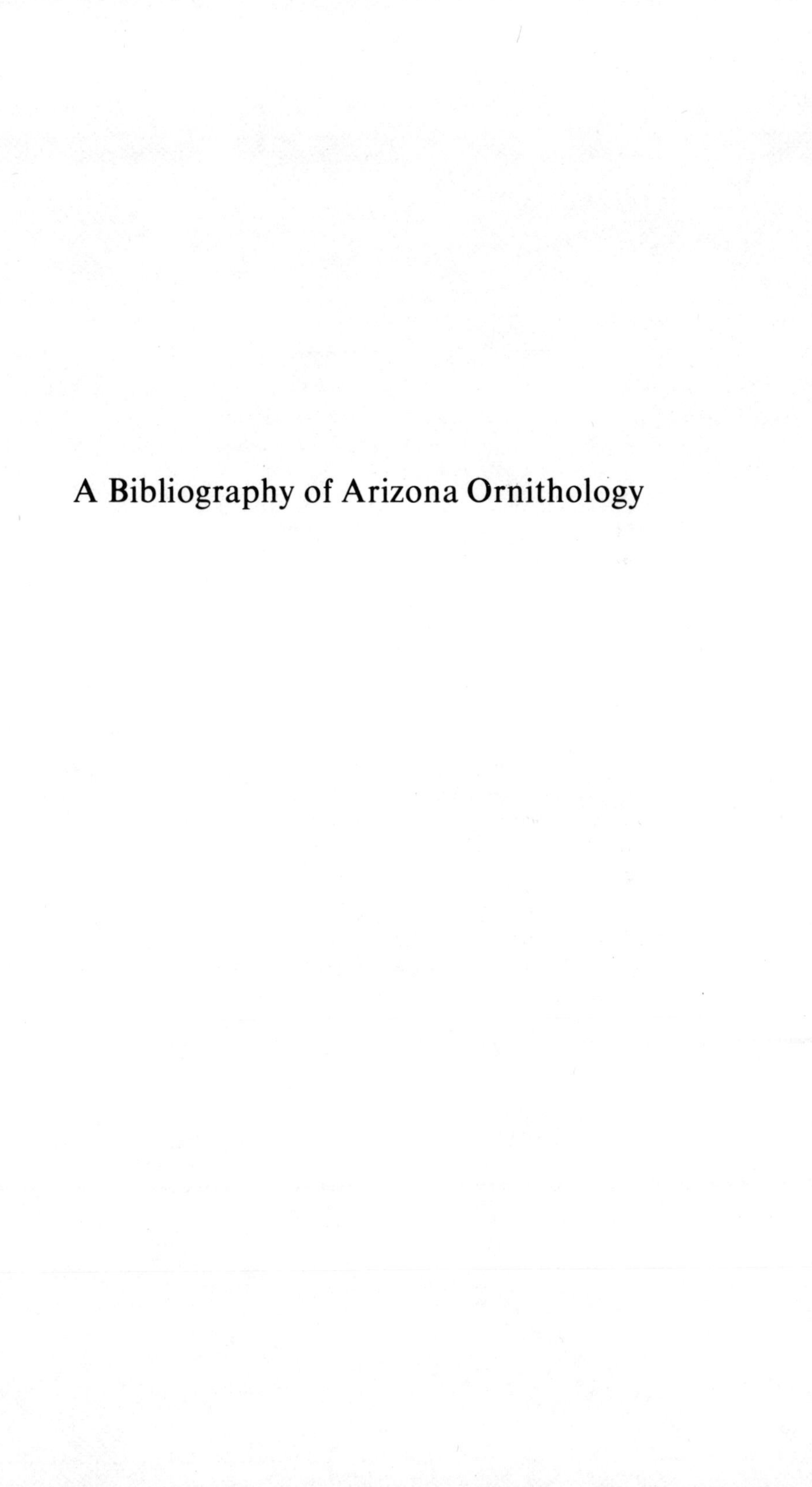

A Bibliography of Arizona Ornithology

1. Nomenclature and Taxonomy

The entries in this section consist chiefly of descriptions of new subspecies of birds whose distribution is confined wholly or in part to Arizona. A few relate to nomenclatural changes and to discussions of observed hybrids in the state.

Aldrich, J. W.

1946. New subspecies of birds from western North America. Proc. Biol. Soc. Wash. 59: 129–135.

Dumetella carolinensis ruficrissa and *Spinus pinus vagans* are listed from Arizona.

———

1951. A review of the races of the Traill's Flycatcher. Wilson Bull. 63: 192–197.

The race in the desert region of Arizona is *Empidonax traillii extimus*.

———

1968. Population characteristics and nomenclature of the Hermit Thrush. Proc. U.S. Nat. Mus. 124, no. 3637, 33 pp.

More changes in subspecific names, some affecting Arizona migrants.

——— and K. P. Baer

1970. Status and speciation in the Mexican Duck *(Anas diazi)*. Wilson Bull. 81: 63–73.

Anas diazi novimexicana is not considered a valid subspecies.

Alexander, G.

1945. Natural hybrids between *Dendroica coronata* and *D. auduboni*. Auk 62: 623–626.

Discusses Arizona hybrids mentioned by Monson and Phillips in Condor 43: 108–112; 1941.

Allen, J. A.

1886. The type specimen of *Colinus ridgwayi*. Auk 3: 483.

Correction of an error in Bull. Am. Mus. Nat. Hist. 1 (7): 276. The type specimen is in the collection of G. F. Morcom of Chicago.

———

1887. A further note on *Colinus ridgwayi*. Auk 4: 74–75.

Correction of an error and a description of another specimen from Arizona.

Allen, J. A.
1889. Note on the first plumage of *Colinus ridgwayi.* Auk 6: 189.
Description of a specimen taken "near Tubal [Tubac?], seventy miles south of Tucson."

———
1892. The North American species of the genus *Colaptes,* considered with special reference to the relationships of *C. auratus* and *C. cafer.* Bull. Am. Mus. Nat. Hist. 4 (2): 21–44.
Discusses variation and hybridization; lists 82 specimens from Arizona.

Allen, R. W.
1950. Notes on some winter birds of north central New Mexico. Auk 67: 252–253.
J. W. Aldrich regards *Corvus brachyrhynchos hargravei* as a valid subspecies; specimens from Arizona examined.

Anthony, A. W.
1894. Notes on the genus *Heleodytes,* with a description of a new subspecies. Auk 11: 210–214.
Comparisons made with Arizona specimens of *H. brunneicapillus.*

———
1895. New races of *Colaptes* and *Passerella* from the Pacific Coast. Auk 12: 347–349.
Includes remarks on *C. chrysoides* from Arizona.

Arvey, M. D.
1951. Phylogeny of the Waxwings and allied birds. Univ. Kansas Publ. Mus. Nat. Hist. 3 (3): 473–530.
Arizona specimens of *Phainopepla* are figured.

Bailey, A.M.
1928. A study of the Snowy Herons of the United States. Auk 45: 430–440.
Western birds are considered to be *Egretta thula brewsteri;* Arizona specimens are given.

Baird, S. F.
1854. Descriptions of new birds, collected between Albuquerque, N.M., and San Francisco, California, during the winter of 1853–54, by Dr. C. B. R. Kennerly and H. B. Möllhausen, naturalists attached to the survey of the Pacific R. R. Route, under Lt. A. W. Whipple. Proc. Acad. Nat. Sci. Phila. 7:118–120.
Seven new species are described from Arizona localities.

——— and R. Ridgway
1873. On some new forms of American birds. Bull. Essex Inst. 5 (12): 197–201.
Includes description of *Cyanocitta ultramarina* var. *Arizonae* Ridgway, type from Fort Buchanan, Arizona.

Bangs, O.
1905. Descriptions of seven new subspecies of American birds. Proc. Biol. Soc. Wash. 18: 151–156.
Includes mention of *Scardafella inca* in southern Arizona. No geographical variation found.

———

1907. A new race of the Hepatic Tanager. Proc. Biol. Soc. Wash. 20: 29–30.
P. h. dextra of eastern Mexico is described; Arizona birds are found to be larger than those of eastern Mexico.

———

1914. The geographic races of the Scaled Quail. Proc. New England Zool. Club 4: 99–100.
The Arizona race is named *Callipepla squamata pallida.*

———

1915. The American forms of *Gallinula chloropus* (Linn.). Proc. New England Zool. Club 5: 93–99.
Includes first description of *G. c. cachinnans.*

———

1927. *Atthis heloisa morcomi* Ridgway, not a valid subspecies. Condor 29: 118–119.
The Arizona specimens are regarded as *A. h. heloisa.*

———

1930. Types of birds now in the Museum of Comparative Zoölogy. Bull. Mus. Comp. Zoöl. 70 (4): 144–426.
Includes several species from Arizona.

——— and G. K. Noble

1918. List of birds collected on the Harvard Peruvian expedition of 1916. Auk 35: 442–463.
The Arizona race of the Caracara is named *Polyborus cheriway auduboni.*

——— and T. E. Penard

1921. Descriptions of six new subspecies of American birds. Proc. Biol. Soc. Wash. 34: 89–92.
Describes *Nuttallornis borealis majorinus,* with specimens from Arizona.

Banks, R. C.

1970. Molt and taxonomy of Red-breasted Nuthatches. Wilson Bull. 82: 201–205.
Agrees with Phillips, Marshall, and Monson (1964: 114) that *Sitta canadensis clariterga* of Burleigh, 1960, is not a valid subspecies.

——— and N. K. Johnson

1961. A review of North American hybrid Hummingbirds. Condor 63: 3–28.
Includes a discussion of *Calypte costae* x *Selasphorus platycercus* from the Rincon Mountains, Arizona.

Behle, W. H.

1942. Notes on the synonymy and distribution of the Horned Larks of Utah. Proc. Utah Acad. Sci. Arts and Letters 19 and 20: 153–156.
Comparisons made with Arizona subspecies.

———

1942. Distribution and variation of the Horned Larks *(Otocoris alpestris)* of western North America. Univ. Calif. Publ. Zool. 46 (3): 205–316.
Includes three Arizona races.

Behle, W. H.

1948. Systematic comment on some geographically variable birds occurring in Utah. Condor 50: 71–80.

Remarks on *Empidonax difficilis, Psaltriparus minimus,* and *Hylocichla guttata* from Arizona.

———

1950. Clines in the Yellow-throats of western North America. Condor 52: 193–219.

Characters, measurements, and maps of distribution of Arizona races are given.

———

1950. A new race of Mountain Chickadee from the Utah-Idaho area. Condor 52: 273–274.

P. g. wasatchensis intergrades with *P. g. gambeli* in northern Arizona.

———

1951. A new race of the Black-capped Chickadee from the Rocky Mountain region. Auk 68: 75–79.

Parus atricapillus garrinus, new subspecies, is "probably" the form recorded from northern Arizona.

———

1956. A systematic review of the Mountain Chickadee. Condor 58: 51–70.

Includes racial characters, map of distribution, and list of specimens of *Parus gambeli gambeli* of Arizona.

———

1968. A new race of the Purple Martin from Utah. Condor 70: 166–169.

Progne subis arboricola extends into the mountains of northern and south-eastern Arizona.

——— and J. W. Aldrich

1947. Description of a new Yellowthroat *(Geothlypis trichas)* from the northern Rocky Mountain-Great Plains region. Proc. Biol. Soc. Wash. 60: 69–72.

Geothlypis t. campicola occurs in Arizona in migration.

Bishop, L. B.

1915. Description of a new race of Savannah Sparrow and suggestions on some California birds. Condor 17: 185–189.

Discusses specimens of *Zamelodia melanocephala capitalis* from Arizona.

———

1921. Description of a new Loon. Auk 38: 364–370.

Gavia immer elasson, new subspecies; a specimen from the Colorado River, California, is listed.

———

1931. Sexual dichromatism in the Pygmy Owl. Proc. Biol. Soc. Wash. 44: 97–98.

Mentions Arizona specimens.

———

1933. Two apparently unrecognized races of North American birds. Proc. Biol. Soc. Wash. 46: 201–206.

Hylocichla guttata dwighti migrates through Arizona. Arizona races of *Lanius ludovicianus* are discussed.

———

1938. An apparently unrecognized race of Redwing from Utah. Trans. San Diego Soc. Nat. Hist. 9 (1): 1–4.

A. p. utahensis winters in Arizona. Includes specimens of *A. p. utahensis, nevadensis,* and *sonoriensis,* in Arizona with measurements.

Blake, E. R.

1950. Report on a collection of birds from Guerrero, Mexico. Fieldiana 31: 375–393.

Includes comments on specimens of *Accipiter gentilis apache* from Arizona.

Bond, G. M.

1963. Geographic variation in the Thrush *Hylocichla ustulata.* Proc. U.S. Nat. Mus. 114 (3471): 373–387.

H. u. oedica is a migrant at Fort Huachuca.

Bond, R. M.

1943. Variation in western Sparrow Hawks. Condor 45: 168–185.

Arizona birds are considered intergrades.

Brandt, H.

1945. A new wren from Arizona. Auk 62: 574–577.

Troglodytes brunneicollis vorhiesi, type from the Huachuca Mountains.

Breninger, G. F.

1898. Hybridization of Flickers. Osprey 3: 13.

Several supposed hybrids from southern Arizona.

Brewster, W.

1881. On the affinities of certain *Polioptilae,* with a description of a new species. Bull. Nuttall Ornith. Club 6: 101–107.

Polioptila plumbea and *P. melanura* are believed to be the same species; Arizona specimens are mentioned.

———

1887. Three new forms of North American birds. Auk 4: 145–149.

Includes first description of *Phalaenoptilus nuttalli nitidus,* with specimens from Arizona.

———

1888. On three apparently new subspecies of Mexican birds. Auk 5: 136–139.

Describes *Dendroica aestiva sonorana;* range includes southern Arizona.

———

1890. A new subspecies of the Solitary Sandpiper. Auk 7: 377–379.

Totanus solitarius cinnamomeus; specimens from Fort Verde, Arizona.

Brodkorb, P.

1935. Two new subspecies of the Red-shafted Flicker. Occas. Papers Mus. Zool. Univ. Mich. 13 (314): 1–3.

Colaptes cafer canescens and *C. c. chihuahuae,* both in Arizona.

Brodkorb, P.

1936. Geographical variation in the Pinon Jay. Occas. Papers Mus. Zool. Univ. Mich. 13 (332): 1–3.

Gymnorhinus cyanocephalus cassinii is assigned to Arizona.

———

1949. Variation in the North American forms of Western Flycatcher. Condor 51: 35–39.

A critical discussion; *E. d. difficilis* and *E. d. hellmayri* are recognized as breeding birds of Arizona.

Brooks, A.

1937. A nondescript Blackbird from Arizona. Auk 54: 160–163.

Of doubtful origin and left unnamed.

Bryant, W. E.

1888. Description of a new subspecies of Song Sparrow from Lower California, Mexico. Proc. Calif. Acad. Sci. 2nd ser. 1: 197–200.

Includes measurements of *Melospiza fasciata fallax* from Tucson.

———

1890. Notices of supposed new birds. Zoe 1: 148–150.

Proposes *Auriparus flaviceps ornatus* for the western paler Verdins. Tucson specimens mentioned.

Burleigh, T. D.

1960. Three new subspecies of birds from western North America. Auk 77: 210–215.

Sitta canadensis clariterga described as new; range includes "southern Arizona."

———

1960. Geographic variation in the Western Wood Pewee. Proc. Biol. Soc. Wash. 73: 141–146.

Arizona birds are described as *Contopus sordidulus veliei;* many specimens listed.

——— and G. H. Lowery, Jr.

1940. Birds of the Guadalupe Mountain region of western Texas. Occas. Papers Mus. Zool. Louisiana State Univ. 8: 85–151.

Includes remarks on type of *Vireo solitarius plumbeus* from Fort Whipple, Arizona.

——— and G. H. Lowery, Jr.

1942. Notes on the birds of southeastern Coahuila. Occas. Papers Mus. Zool. Louisiana State Univ. 12: 185–212.

Arizona specimens of *Toxostoma dorsale* are compared.

Chapman, F. M.

1897. Preliminary descriptions of new birds from Mexico and Arizona. Auk 14: 310–311.

First description of *Contopus pertinax pallidiventris,* type from Pima County, Arizona.

———

1900. A study of the genus *Sturnella.* Bull. Am. Mus. Nat. Hist. 13 (22): 297–320.

Arizona specimens of *S. magna neglecta* are figured.

———

1901. A new race of the Great Blue Heron, with remarks on the status and range of *Ardea wardi.* Bull. Am. Mus. Nat. Hist. 14 (8): 87–90.
Mentions specimens from Fort Verde.

———

1904. A new Grouse from California. Bull. Am. Mus. Nat. Hist. 20 (11): 159–162.
Mentions specimens of *Dendragapus obscurus obscurus* from Arizona.

Coale, H. K.
1914. San Lucas Verdin *(Auriparus flaviceps lamprocephalus)* in California. Auk 31: 543.
Measurements given of *A. flaviceps* from Cochise County, Arizona.

———

1923. A new subspecies of the Little Black Rail. Auk 40: 88–90.
Cresciscus jamaicensis stoddardi, range "west . . . to Arizona."

Conover, B.
1944. The races of the Solitary Sandpiper. Auk 61: 537–544.
Specimens of *T. s. cinnamomea* listed from Tucson.

Cooper, J. G.
1861. New California animals. Proc. Calif. Acad. Sci. 2: 118–123.
Includes original descriptions of *Athene whitneyi* and *Helminthophila luciae.*

Coues, E.
1872. Studies of the Tyrannidae. Part I. Revision of the species of *Myiarchus.* Proc. Acad. Nat. Sci. Phila.: 56–81.
Includes remarks on a specimen of *M. cinerascens* from Arizona.

———

1873. Some United States birds, new to science, and other things ornithological. Amer. Nat. 7: 321–331.
Includes descriptions of Rufous-winged Sparrow and Bendire's Thrasher.

———

1881. A curious *Colaptes.* Bull. Nuttall Ornith. Club 6: 183.
A hybrid of *C. mexicanus* and *auratus* from Fort Whipple, Arizona.

Court, E. J.
1908. Treganza Blue Heron. Auk 25: 291–296.
First description of *Ardea herodias treganzai;* specimen from Fort Lowell, Arizona.

Davis, J.
1965. Natural history, variation, and distribution of the Strickland's Woodpecker. Auk 82: 537–590.
An important critical paper; the Arizona Woodpecker is considered only subspecifically distinct from *stricklandi.*

Delacour, J.
1941. On the species of *Otus scops.* Zoologica 26: 133–142.
Contains distribution, wing formula, dimensions of wings, and color phases of *Otus scops flammeolus;* text figure of specimen of wing of bird from the Huachuca Mountains.

Dickerman, R. W.
1964. Notes on the Horned Larks of western Minnesota and the Great Plains. Auk 81: 430–432.
Phillips and Marshall believe that the western birds should be called *E. a. occidentalis*.

——— and A. R. Phillips
1966. A new subspecies of the Boat-tailed Grackle from Mexico. Wilson Bull. 78: 129–131.
Includes measurements of *C. m. monsoni,* presumably from Arizona, but no localities are given.

Dickey, D. R.
1923. Description of a new Clapper Rail from the Colorado River Valley. Auk 40: 90–94.
Rallus yumanensis, summer resident along the Colorado River above Yuma.

———
1928. A new Poor-will from the Colorado River Valley. Condor 30: 152–153.
Phalaenoptilus nuttallii hueyi, new subspecies from the lower Colorado Valley in Arizona and California.

——— and A. J. van Rossem
1925. A new Red-winged Blackbird from western Mexico. Proc. Biol. Soc. Wash. 38: 131–132.
Includes measurements of *A. p. sonoriensis* from Arizona.

Duvall, A. J.
1945. Distribution and taxonomy of the Black-capped Chickadees of North America. Auk 62: 49–69.
P. a. nevadensis from "central-northern Arizona."

———
1945. Variation in *Carpodacus purpureus* and *Carpodacus cassinii.* Condor 47: 202–205.
C. p. rubidus, new subspecies, occurs at Fort Verde, Arizona, in winter; range of *C. c. cassinii* includes Arizona.

Dwight, J., Jr.
1890. The Horned Larks of North America. Auk 7: 138–158.
Contains original description of *Otocoris alpestris adusta,* type locality Fort Huachuca, Arizona.

——— and L. Griscom
1927. A revision of the geographical races of the Blue Grosbeak *(Guiraca caerulea).* Amer. Mus. Novit. 257: 1–5.
Arizona birds are *G. c. interfusa,* type from Fort Lowell.

Ficken, M. S. and R. W. Ficken
1965. Comparative ethology of the Chestnut-sided Warbler, Yellow Warbler, and American Redstart. Wilson Bull. 77: 363–375.
Setophaga picta is placed in *Myioborus.*

Figgins, J. D.
1930. Proposals relative to certain subspecific groups of *Carpodacus mexicanus.* Proc. Colo. Mus. Nat. Hist. 9:1–3.
Eastern Arizona birds should be called *C. m. obscurus.*

Fisher, W. K.
1902. Status of *Cyanocitta stelleri carbonacea* Grinnell. Condor 4: 41–44.
Includes *C. s. diademata* of Arizona in the key to the western races.

Friedmann, H.
1933. Critical notes on American Vultures. Proc. Biol. Soc. Wash. 46: 187–190.
The western Turkey Vulture is named *Cathartes aura teter;* its range includes Arizona.

Gambel, W.
1843. Descriptions of some new and rare birds of the Rocky Mountains and California. Proc. Acad. Nat. Sci. Phila. 1: 259–262.
No definite localities in Arizona are listed.

George, W. G.
1962. The classification of the Olive Warbler, *Peucedramus taeniatus.* Amer. Mus. Novit. 2103: 1–41.
Data on distribution, nesting, and song in Arizona. The species is tentatively assigned to the Muscicapidae.

Grinnell, G. B.
1884. A Quail new to the United States fauna. Forest and Stream 22: 243.
Identified by Ridgway as *Ortyx graysoni.*

Grinnell, J.
1908. The southern California Chickadee. Condor 10: 29–30.
Arizona specimens are *Parus gambeli gambeli.*

———
1908. The name of the California Least Vireo. Auk 25: 85–86.
The Arizona Least Vireo should be *Vireo bellii pusillus.*

———
1909. Three new Song Sparrows from California. Univ. Calif. Publ. Zool. 5 (3): 265–269.
Describes *Melospiza melodia saltonis;* range includes the lower Colorado River Valley.

———
1909. A new Cowbird of the genus *Molothrus* with a note on the probable genetic relationships of the North American forms. Univ. Calif. Publ. Zool. 5 (5): 275–281.
Includes measurements of *Molothrus ater obscurus* from Arizona.

———
1911. Description of a new Spotted Towhee from the Great Basin. Univ. Calif. Publ. Zool. 7 (8): 309–311.
Pipilo maculatus curtatus winters in the lower Colorado River Valley.

———
1917. The subspecies of *Hesperiphona vespertina.* Condor 19: 17–22.
H. v. warreni, new subspecies, north-central Arizona; *H. v. montana,* from Huachuca and Chiricahua Mountains.

———
1918. The subspecies of the Mountain Chickadee. Univ. Calif. Publ. Zool. 17 (17): 505–515.
Specimens of *P. g. gambeli* from Arizona were examined.

Grinnell, J.

1926. A critical inspection of the Gnatcatchers of the Californias. Proc. Calif. Acad. Sci. 15 (16): 493–500.

Polioptila caerulea amoenissima is proposed for the western form. *P. m. melanura* specimens from Arizona are discussed.

———

1927. Six new subspecies of birds from Lower California. Auk 44: 67–72.

Hooded Orioles of Arizona, southern California, and northern Lower California apparently migrate southeastward into the mainland of Mexico.

———

1927. A new race of Crissal Thrasher, from northwestern Lower California. Condor 29: 127.

Compares specimens of *T. c. crissale* from Arizona.

———

1928. Notes on the systematics of west American birds. I. Condor 30: 121–124.

Comments on *Spinus pinus macropterus* and *Progne subis subis* in Arizona.

———

1931. The type locality of the Verdin. Condor 33: 163–168.

The eastern Arizona form is called *Auriparus flaviceps ornatus;* western birds are given a new name, *A. f. acaciarum.*

———

1934. The New Mexico race of Plain Titmouse. Condor 36: 251–252.

Description of *Baeolophus inornatus plumbescens;* range includes parts of Arizona south of the Colorado and Little Colorado rivers.

——— and Wm. H. Behle

1935. Comments upon the subspecies of *Catherpes mexicanus.* Condor 37: 247–251.

Disagree with Oberholser and find that *C. m. conspersus* is the only recognizable form in the Rocky Mountain region.

——— and H. S. Swarth

1926. Geographic variation in *Spizella atrogularis.* Auk 43: 475–478.

Lists specimens of *S. a. atrogularis* from Arizona.

Griscom, L.

1929. Studies from the Dwight collection of Guatemala birds. I. Amer. Mus. Novit. 379: 1–13.

Arizona birds are referred to the newly described subspecies *Hylocharis leucotis borealis.*

———

1929. Notes on the Rough-winged Swallow (*Stelgidopteryx serripennis* (Aud.)) and its allies. Proc. New Eng. Zool. Club 11: 67–72.

Specimens from southern Arizona are *S. ruficollis serripennis.*

———

1934. The ornithology of Guerrero, Mexico. Bull. Mus. Comp. Zool. 75 (10): 367–422.

Contains a revision of *Amazilia violiceps; Cyanomyia salvini* is regarded as a hybrid. Comments on several other Arizona species.

Hardy, J. W.
1969. A taxonomic revision of the New World jays. Condor 71: 360–375.
Includes sound spectrograph of Steller Jay at Mt. Lemmon, Arizona.

Hargitt, E.
1886. Notes on Woodpeckers. No. XI. On a new species from Arizona. Ibis, 5th ser. 4 (14): 112–115.
First description of *Picus arizonae*, type from the Santa Rita Mountains, Arizona.

Hawbecker, A. C.
1948. Analysis of variation in western races of the White-breasted Nuthatch. Condor 50: 26–39.
Map of distribution of *S. c. nelsoni* in Arizona; many specimens listed.

Heermann, A. L.
1854. Additions to North American ornithology, with descriptions of new species of the genera *Actidurus, Podiceps*, and *Podylymbus*. Proc. Acad. Nat. Sci. Phila. 7: 177–180.
Includes description of *Hypotriorchis aurantius*, observed on the "vast plains of New Mexico."

Henshaw, H. W.
1884. Description of a new Song Sparrow from the southern border of the United States. Auk 1: 223–224.
Melospiza fasciata fallax is the breeding form in southern Arizona; *M. f. montana* is a winter visitant at Tucson.

Howe, R. H., Jr.
1900. Ranges of *Hylocichla fuscescens* and *Hylocichla fuscescens salicicola* in North America. Auk 17: 18–21.
"south to . . . Arizona." given under *H. f. salicicola.*

Huey, L. M.
1944. A hybrid Costa's x Broad-tailed Hummingbird. Auk 61: 636–637.
From the Rincon Mountains, Arizona.

Jenks, R.
1936. A new race of Golden-crowned Kinglet from Arizona. Condor 38: 239–244.
Regulus regulus apache from the White Mountains.

———
1938. A new subspecies of Pine Grosbeak from Arizona with critical notes on other races. Condor 40: 28–35.
Pinicola enucleator jacoti, type from White Mountains.

Johnsgard, P. A.
1961. Evolutionary relationships among the North American Mallards. Auk 78: 3–43.
A. d. novimexicana is considered a synonym of *diazi;* map shows a specimen from southeastern Arizona.

———
1970. A summary of intergeneric New World Quail hybrids, and a new intergeneric hybrid combination. Condor 72: 85–88.
Lophortyx and *Callipepla* should be merged with *Colinus.*

Johnson, N. K.

1963. Biosystematics of sibling species of Flycatchers in the *Empidonax hammondii-oberholseri-wrightii* complex. Univ. Calif. Publ. Zool. 66 (2): 79–238.

A critical review. Map shows *E. wrightii* and *E. oberholseri* breeding in Arizona. Arizona specimens listed.

———

1969. [Review of] Hybridization in the Flickers *(Colaptes)* of North America. Bull. Amer. Mus. Nat. Hist. 129: 307–428, 1965. Variation in West Indian Flickers *(Aves, Colaptes)*. Bull. Florida State Mus. 10: 1–42, 1965. Variation in Central American Flickers. Wilson Bull. 79: 5–21, 1967. All by L. L. Short, Jr. Wilson Bull. 81: 225–230.

A good, critical review.

Johnston, D. W.

1961. The biosystematics of American Crows. Univ. Wash. Press. Seattle: viii + 119 pp.

C. b. hargravei is not recognized.

Johnston, R. F. and R. K. Selander

1964. House Sparrows: rapid evolution of races in North America. Science 144 (3618): 548–550.

Arizona specimens are found to be pale.

Kennard, F. H.

1919. Notes on a new subspecies of Blue-winged Teal. Auk 36: 455–460.

Querquedula discors albinucha breeds "possibly as far west as Arizona. . . ." Both races, however, have been collected in Arizona.

Lanyon, W. E.

1960. Relationship of the House Wren *(Troglodytes aedon)* of North America and the Brown-throated Wren *(Troglodytes brunneicollis)* of Mexico. [In] Proc. 12th Int. Ornith. Cong., Helsinki, 1958: 450–458.

They are considered conspecific.

———

1961. Specific limits and distribution of Ash-throated and Nutting Flycatchers. Condor 63: 421–449.

Includes discussion of Arizona birds.

———

1962. Specific limits and distribution of Meadowlarks of the desert grassland. Auk 79: 183–207.

Extensive discussion of morphological, vocal, and ecological differences in the two species of Arizona Meadowlarks.

Lawrence, G. N.

1852. Description of new species of birds, of the genera *Toxostoma* Wagler, *Tyrannula* Swainson, and *Plectrophanes* Meyer. Ann. Lyc. Nat. Hist. N.Y. 5: 121–123.

Toxostoma Le Contei, from California near the junction of the Gila and Colorado rivers.

———
1877. Note on *Doricha enicura* (Vieill.). Bull. Nuttall Ornith. Club 2: 108–109.
Reexamination shows it to be a female of *Calothorax lucifer*.

Linsdale, J. M.
1938. Geographic variation in some birds in Nevada. Condor 40: 36–38.
First description of *Penthestes atricapillus nevadensis;* a migrant taken in Betatakin Canyon, Arizona. (See Wetherill, Condor 39: 86, 1937.)

Marshall, J. T., Jr.
1942. *Melospiza melodia virginis* a synonym of *Melospiza melodia fallax*. Condor 44: 233.

——— and W. H. Behle
1942. The Song Sparrows of the Virgin River valley, Utah. Condor 44: 122–124.
M. m. virginis, new subspecies, winters along the lower Colorado River.

McCabe, T. T. and E. B. McCabe
1932. Preliminary studies of western Hermit Thrushes. Condor 34: 26–40.
Arizona specimens are included in map.

——— and A. H. Miller
1933. Geographic variation in the Northern Water-thrushes. Condor 35: 192–197.
Specimens of *S. n. noveboracensis* from Arizona are listed.

Mearns, E. A.
1890. Descriptions of a new species and three new subspecies of birds from Arizona. Auk 7: 243–251.
Junco ridgwayi, Spinus tristis pallidus, Coccothraustes vespertina montana (redescribed), and *Melanerpes formicivorus aculeatus.*

———
1892. A study of the Sparrow Hawks (subgenus *Tinnunculus*) of America, with especial reference to the continental species (*Falco sparverius* Linn.). Auk 9: 252–270.
Original description of *Falco sparverius deserticola,* type from Fort Verde, Arizona.

———
1895. Description of a new heron *(Ardea virescens anthonyi)* from the arid region of the interior of North America. Auk 12: 257–259.
Breeds on the Verde River and along the Mexican boundary line.

———
1902. Descriptions of three new birds from the southern United States. Proc. U.S. Nat. Mus. 24: 915–926.
Includes original description of *Sitta carolinensis nelsoni;* type from Huachuca Mountains, Arizona.

———
1902. Description of a new Swallow from the western United States. Proc. Biol. Soc. Wash. 15: 31–32.
Tachycineta lepida; range includes Arizona.

Mearns, E. A.
1902. The Cactus Wrens of the United States. Auk 19: 141–145.
First description of *Heleodytes brunneicapillus anthonyi;* type from Adonde Siding, Arizona.

Miller, A.H.
1930. Two new races of the Loggerhead Shrike from western North America. Condor 32: 155–156.
Describes *L. l. sonoriensis,* type from the Chiricahua Mountains; and *L. l. nevadensis,* whose range extends into northern Arizona.

———
1931. Systematic revision and natural history of the American Shrikes *(Lanius).* Univ. Calif. Publ. Zool. 38: 11–242.
Includes Arizona subspecies.

———
1938. Hybridization of Juncos in captivity. Condor 40: 92–93.
A male *Junco caniceps dorsalis* taken from Flagstaff to Berkeley, California, mates with a *Junco oreganus pinosus* and hatches normal young.

———
1939. Analysis of some hybrid populations of Juncos. Condor 41: 211–214.
Discusses Arizona hybrids.

———
1941. Speciation in the avian genus *Junco.* Univ. Calif. Publ. Zool. 44 (3): 173–434.
This exhaustive study includes all Arizona races.

———
1941. A review of centers of differentiation for birds in the western Great Basin region. Condor 43: 257–267.
Includes a comparison of *Chondestes grammacus actitus* with Arizona specimens of *C. g. strigatus.*

———
1948. Further observations on variation in Canyon Wrens. Condor 50: 83–85.
Variation, individual and local, is found to be great; the names *punctulatus, griseus,* and *polioptilus* are discarded in favor of one name *C. m. conspersus.*

———
1949. Some concepts of hybridization and intergradation in wild populations of birds. Auk 66: 338–342.
Includes a discussion of the Juncos of the Kaibab Plateau, Arizona.

———
1955. A hybrid Woodpecker and its significance in speciation in the genus *Dendrocopus.* Evolution 9: 317–321.
Brief mention of habitats of *D. scalaris* and *D. villosus* in Arizona.

———
1955. The avifauna of the Sierra del Carmen of Coahuila, Mexico. Condor 57: 154–178.
Includes a few comparisons with Arizona specimens.

——— and T. T. McCabe

1935. Racial differentiation in *Passerella (Melospiza) lincolnii*. Condor 37: 144–160.

P. l. alticola, new subspecies, specimen from the White Mountains, Arizona; maps of winter distribution of races.

——— and L. Miller

1951. Geographic variation of the Screech Owls of the deserts of western North America. Condor 53: 161–177.

Otus asio yumanensis, new subspecies from the lower Colorado River Valley; extensive discussion of Arizona specimens.

——— and R. W. Storer

1950. A new race of *Parus sclateri* from the Sierra Madre del Sur of Mexico. Jour. Wash. Acad. Sci. 40: 301–302.

P. s. eidos of Arizona is considered a recognizable race.

Miller, W. DeW.

1906. List of birds collected in northwestern Durango, Mexico, by J. H. Batty, during 1903. Bull. Amer. Mus. Nat. Hist. 22 (10): 161–183.

Includes critical comments on a number of Arizona species.

——— and L. Griscom

1925. Notes on Central American birds, with descriptions of new forms. Amer. Mus. Novit. 183: 1–14.

Peucedramus olivaceus arizonae, new subspecies, described from Arizona.

Moore, R. T.

1932. A new race of *Aimophila carpalis* from Mexico. Proc. Biol. Soc. Wash. 45: 231–234.

Includes measurements of Arizona birds.

———

1937. New races of the genus *Otus* from northwestern Mexico. Proc. Biol. Soc. Wash. 50: 63–68.

Specimens of *Otus asio cineraceus* and *O. a. gilmani* from Arizona examined.

———

1939. Two new races of *Carpodacus mexicanus*. Proc. Biol. Soc. Wash. 52: 105–112.

Remarks on Arizona specimens.

———

1939. The Arizona Broad-billed Hummingbird. Auk 56: 313–319.

Arizona birds are *C. l. magicus*.

———

1939. A review of the House Finches of the subgenus *Burrica*. Condor 41: 177–205.

Maps of distribution, characters of subspecies, and lists of specimens examined from Arizona.

———

1940. New races of *Empidonax* from Middle America. Proc. Biol. Soc. Wash. 53: 23–29.

E. d. immodulatus, new subspecies, extends north to Santa Rita Mountains, Arizona.

Moore, R. T.

1940. Notes on Middle American *Empidonaces.* Auk 57: 349–389.
Critical review, includes *E. d. difficilis, E. d. immodulatus,* and *E. fulvifrons pygmaeus* specimens from Arizona.

———

1941. Notes on *Toxostoma curvirostre* of Mexico, with description of a new race. Proc. Biol. Soc. Wash. 54: 211–216.
T. c. celsum, new subspecies from southeastern Arizona.

———

1946. The status of *Dendroica auduboni nigrifrons* in the United States. Auk 63: 241–242.
Arizona specimens are allocated to *D. a. memorabilis.*

——— and J. L. Peters

1939. The genus *Otus* of Mexico and Central America. Auk 56: 38–56.
Critical review. Includes *O. a. cineraceus, O. trichopsis aspersus,* and *O. flammeolus flammeolus* in Arizona.

Nelson, E. W.

1884. Brief diagnoses of two new races of North American birds. Auk 1: 165–166.
Includes description of *Astur atricapillus henshawi;* range, Pacific Coast region from southern Arizona to Sitka, Alaska.

———

1898. Description of new birds from Mexico, with a revision of the genus *Dactylortyx.* Proc. Biol. Soc. Wash. 12: 57–68.
Includes remarks on *Heleodytes brunneicapillus* of the southwestern United States.

———

1899. Descriptions of new birds from northwestern Mexico. Proc. Biol. Soc. Wash. 13: 25–31.
Includes measurements of *Cardinalis cardinalis superbus* from Tucson.

———

1899. Descriptions of new birds from Mexico. Auk 16: 25–31.
Mentions that birds from southern Arizona are typical *Callipepla gambeli.*

———

1900. Description of a new subspecies of *Meleagris gallopavo* and proposed changes in the nomenclature of certain North American birds. Auk 17: 120–126.
M. g. merriami, type from 47 miles southwest of Winslow, Arizona.

———

1900. Descriptions of thirty new North American birds, in the Biological Survey collection. Auk 17: 253–270.
Original description of *Cyrtonyx montezumae mearnsi,* type from Fort Huachuca, Arizona.

———

1904. A revision of the North American mainland species of *Myiarchus.* Proc. Biol. Soc. Wash. 17: 21–50.
Includes all Arizona members of the genus.

———

1905. Notes on the names of certain North American birds. Proc. Biol. Soc. Wash. 18: 121–126.

Cathartes aura septentrionalis is proposed for the birds of the United States, including the "southern border."

———

1910. A new subspecies of Pigmy Owl. Proc. Biol. Soc. Wash. 23: 103–104.

Glaucidium gnoma pinicola, range includes Arizona.

Oberholser, H. C.

1896. Critical remarks on the Mexican forms of the genus *Certhia*. Auk 13: 314–318.

The Arizona race is named *Certhia familiaris albescens*.

———

1897. Critical remarks on *Cistothorus palustris* (Wils.) and its western allies. Auk 14: 186–196.

Description of *C. p. plesius*, listing specimens from Arizona.

———

1897. Description of a new *Empidonax*, with notes on *Empidonax difficilis*. Auk 14: 300–303.

Discusses *E. difficilis* from the Santa Catalina Mountains, Arizona.

———

1897. Critical notes on the genus *Auriparus*. Auk 14: 390–394.

Specimens of *A. flaviceps* from Texas and southeastern Arizona are similar.

———

1898. A revision of the Wrens of the genus *Thryomanes* Sclater. Proc. U.S. Nat. Mus. 21: 421–450.

T. bewicki eremophilus is the Arizona form. *T. b. drymoecus* "casually to Arizona."

———

1899. Description of a new *Hylocichla*. Auk 16: 23–25.

Hylocichla ustulata oedica. Ranges "south in winter to Arizona. . . ." Specimen from Fort Huachuca.

———

1899. A synopsis of the genus *Contopus* and its allies. Auk 16: 330–337.

Horizopus pertinax pallidiventris is proposed for the Arizona race.

———

1902. A review of the Larks of the genus *Otocoris*. Proc. U.S. Nat. Mus. 24: 801–883.

Arizona subspecies are: *O. alpestris adusta, O. a. leucolaema, O. a. praticola, O. a. occidentalis, O. a. leucansiptila*, and *O. a. aphrasta*.

———

1903. A review of the genus *Catherpes*. Auk 20: 196–198.

First description of *C. mexicanus polioptilus*.

———

1903. The North American forms of *Astragalinus psaltria* (Say). Proc. Biol. Soc. Wash. 16: 113–116.

A. p. hesperophilus, new subspecies; range includes Arizona.

Oberholser, H. C.

1903. A synopsis of the genus *Psaltriparus*. Auk 20: 198–201.
Psaltriparus santaritae is an immature male of *P. melanotis lloydi*.

———

1904. A revision of the American Great Horned Owls. Proc. U.S. Nat. Mus. 27: 177–192.
Asio magellanicus pallescens and *A. m. pacificus* ascribed to Arizona.

———

1904. A review of the Wrens of the genus *Troglodytes*. Proc. U.S. Nat. Mus. 27: 197–210.
T. a. aztecus: western United States.

———

1905. The forms of *Vermivora celata* (Say). Auk 22: 242–247.
First description of *V. c. orestera,* breeding on Mt. Graham, Arizona.

———

1911. A revision of the forms of the Hairy Woodpecker (*Dryobates villosus* [Linnaeus]). Proc. U.S. Nat. Mus. 40: 595–621.
The Arizona forms are *D. v. icastus* and *D. v. leucothorectis*.

———

1911. A revision of the forms of the Ladder-backed Woodpecker (*Dryobates scalaris* [Wagler]). Proc. U.S. Nat. Mus. 41: 139–159.
Describes *D. s. cactophilus,* type from Tucson, Arizona.

———

1912. A revision of the subspecies of the Green Heron (*Butorides virescens* [Linnaeus]). Proc. U.S. Nat. Mus. 42: 529–577.
The Arizona form is *B. v. anthonyi.*

———

1912. A revision of the forms of the Great Blue Heron (*Ardea herodias* Linnaeus). Proc. U.S. Nat. Mus. 43: 531–559.
The Arizona form is *A. h. treganzai.*

———

1914. A monograph of the genus *Chordeiles* Swainson, type of a new family of Goatsuckers. U.S. Nat. Mus. Bull. 86: 1–123.

———

1915. Critical notes on the subspecies of the Spotted Owl, *Strix occidentalis* (Xantus). Proc. U.S. Nat. Mus. 49 (2106): 251–257.
S. o. lucida, new name for the Arizona form.

———

1917. Notes on North American birds. II. Auk 34: 321–329.
Discusses *Vireo bellii arizonae, Baeolophus wollweberi annexus, Vermivora celata orestera,* and *Molothrus ater obscurus.*

———

1917. A synopsis of the races of *Bombycilla garrula* (Linnaeus). Auk 34: 330–333.
B. g. pallidiceps is "Casual in Arizona."

———

1918. Notes on the subspecies of *Numenius americanus* Bechstein. Auk 35: 188–195.

N. a. americanus and *N. a. occidentalis* listed from Arizona.

———

1918. Description of a new subspecies of *Cyanolaemus clemenciae.* Condor 20: 181–182.

C. c. bessophilus, type from Fly Park, Chiricahua Mountains.

———

1918. New light on the status of *Empidonax traillii* (Audubon). Ohio Jour. Sci. 18: 85–98.

E. t. brewsteri, new subspecies; range includes Arizona.

———

1918. The Common Ravens of North America. Ohio Jour. Sci. 18: 213–225.

Corvus corax sinuatus and *C. c. clarionensis* are both assumed to be in southern Arizona.

———

1919. An unrecognized subspecies of *Melanerpes erythrocephalus.* Canadian Field-Nat. 33: 48–50.

The Arizona specimen becomes *M. e. erythrophthalmus.*

———

1919. Notes on the Wrens of the genus *Nannus* Billberg. Proc. U.S. Nat. Mus. 55: 223–236.

N. troglodytes pacificus "winters south to southern Arizona. . . ."

———

1919. Description of a new subspecies of *Piranga hepatica* Swainson. Auk 36: 74–80.

First description of *P. h. oreophasma,* the breeding bird in Arizona.

———

1919. The geographic races of *Hedymeles melanocephalus* Swainson. Auk 36: 408–416.

H. m. papago is the Arizona race.

———

1919. Description of a new subspecies of *Pipilo fuscus.* Condor 21: 210–211.

P. f. albigulus intergrades individually with *P. f. mesoleucus* of Arizona.

———

1919. A revision of the subspecies of *Passerculus rostratus* (Cassin). Ohio Jour. Sci. 19: 344–354.

Includes *P. r. rostratus* in Arizona.

———

1920. A synopsis of the genus *Thryomanes.* Wilson Bull. 32: 18–28.

The Arizona race is *T. bewicki eremophilus,* not *T. b. bairdi.*

———

1921. A revision of the races of *Dendroica auduboni.* Ohio Jour. Sci. 21: 240–248.

First description of *D. a. memorabilis,* range to central and southeastern Arizona.

Oberholser, H. C.
1930. Notes on a collection of birds from Arizona and New Mexico. Sci. Publ. Cleveland Mus. Nat. Hist. 1: 83–124.
Dendroica nigrescens halseii and *Sturnella magna lilianae* are described.

———
1932. Descriptions of new birds from Oregon, chiefly from the Warner Valley region. Sci. Publ. Cleveland Mus. Nat. Hist. 4: 1–12.
Eighteen new subspecies, half of which are listed as present in Arizona.

———
1934. A revision of the North American House Wrens. Ohio Jour. Sci. 34: 86–96.
Arizona specimens of *Troglodytes domesticus parkmanii* are discussed.

———
1937. Descriptions of two new passerine birds from the western United States. Proc. Biol. Soc. Wash. 50: 117–120.
Describes *Cyanocitta stelleri cottami;* range includes northern Arizona.

———
1937. A revision of the Clapper Rails (*Rallus longirostris* Boddaert). Proc. U.S. Nat. Mus. 84: 313–354.
R. l. yumanensis, from the lower Colorado River, but no Arizona specimens listed.

———
1937. Descriptions of three new Screech Owls from the United States. Jour. Wash. Acad. Sci. 27: 354–357.
Otus asio mychophilus, new subspecies from northern Arizona.

———
1942. Description of a new Arizona race of the Grasshopper Sparrow. Proc. Biol. Soc. Wash. 55: 15–16.
Ammodramus savannarum ammolegus, in central southern Arizona.

———
1946. Three new North American birds. Jour. Wash. Acad. Sci. 36: 388–389.
Amphispiza nevadensis campicola winters in southern Arizona.

———
1948. Descriptions of new races of *Geothlypis trichas* (Linnaeus). Privately printed, Cleveland. 4 pp.
Describes 12 new races, 3 of which have been found in Arizona.

Ogilvie-Grant, W. R.
1902. Remarks on the species of American Gallinae recently described, and notes on their nomenclature. Ibis 8th ser. 2: 233–245.
Comments on *Meleagris gallopavo, Lophortyx gambeli,* and *Cyrtonyx montezumae.*

Owen, D. F.
1963. Variation in North American Screech Owls and the subspecies concept. Syst. Zool. 12: 8–14.
Advocates use of name *Otus asio* only, with inclusion of comments on variation in North America. Map shows variation in wing length in North America.

Parkes, K. C.
1950. Further notes on the birds of Camp Barkeley, Texas. Condor 52: 91–93.
Mearns' specimen of *Dichromanassa rufescens* from Camp Verde, Arizona, belongs to the typical race, *rufescens*.

——— and E. R. Blake
1965. Taxonomy and nomenclature of the Bronzed Cowbird. Fieldiana 44: 207–216.
Tangavius aeneus milleri of Arizona is renamed *Molothrus aeneus loyei*.

Paynter, R. A., Jr.
1957. Taxonomic notes on the New World forms of *Troglodytes*. Breviora 71: 1–15.
Includes a discussion of interbreeding of *T. aëdon* and *T. brunneicollis* in Arizona.

Peters, J. L.
1927. Descriptions of new birds. Proc. New England Zool. Club 9: 111–113.
Describes *Penthestes sclateri eidos*, type from the Chiricahua Mountains, Arizona.

——— and L. Griscom
1938. Geographical variation in the Savannah Sparrow. Bull. Mus. Comp. Zool. 80 (13): 445–480.
A specimen of *P. s. alaudinus* from Arizona is listed.

Phillips, A. R.
1939. The type of *Empidonax wrightii* Baird. Auk 56: 311–312.
E. wrightii proves to be *E. griseus;* Wright's Flycatcher is renamed *Empidonax oberholseri*, and the Gray Flycatcher becomes *E. wrightii*.

———
1942. A new Crow from Arizona. Auk 59: 573–575.
C. b. hargravei from the Mogollon Plateau, Arizona, is described.

———
1943. Critical notes on two southwestern sparrows. Auk 60: 242–248.
Melospiza melodia bendirei, new subspecies, type from Tempe Butte, Arizona. *Aimophila botterii arizonae* is proposed for the Arizona birds.

———
1947. The races of MacGillivray's Warbler. Auk 64: 296–300.
Described as new: *O. t. monticola*, type from Hart Prairie, Arizona, breeding; *O. t. austinsmithi* and *O. t. intermedia* are northern races.

———
1948. Geographic variation in *Empidonax traillii*. Auk 65: 507–514.
E. t. extimus, type from Feldman, Arizona, is described; range, southern and western Arizona; *E. t. brewsteri* breeds in northeastern Arizona.

———
1950. The Great-tailed Grackles of the southwest. Condor 52: 78–81.
Describes *Cassidix mexicanus monsoni*, new subspecies, ranging into southeastern Arizona.

Phillips, A. R.

1950. The pale races of the Steller Jay. Condor 52: 252–254.
Describes *Cyanocitta stelleri browni,* new subspecies, type from Carter Canyon, Santa Catalina Mountains, Arizona.

——— 1958. Las subspecies de la Codorniz de Gambel y el problema de los cambios climaticos en Sonora. Univ. Méx. Anales del Inst. de Biol. 29: 361–374.
A critical revision, including Arizona forms.

——— 1959. The nature of avian species. Jour. Ariz. Acad. Sci. 1: 22–30.
A lengthy exposition that includes data on Arizona species.

——— 1964. Notas sistematicas sobre aves Mexicanas. III. Rev. Soc. Mex. Hist. Nat. 25: 217–242.
New subspecies assigned to Arizona: *Cyanocorax coerulescens suttoni, Regulus calendula arizonensis, Pooecetes gramineus altus.*

——— 1966. Further systematic notes on Mexican birds. Bull. British Ornith. Club 86: 148–159.
Describes *Piranga rubra ochracea* from Big Sandy Valley, Mohave County, Arizona.

——— and K. C. Parkes

1955. Taxonomic comments on the Western Wood Pewee. Condor 57: 244–246.
The name becomes *Contopus sordidulus,* and *C. s. placens* is considered a synonym of *C. s. veliei.*

Pitelka, F. A.

1945. Differentiation of the Scrub Jay, *Aphelocoma coerulescens,* in the Great Basin and Arizona. Condor 47: 23–26.
Describes *A. c. nevadae,* new subspecies; range includes Arizona.

——— 1951. Speciation and ecologic distribution in American Jays of the genus *Aphelocoma.* Univ. Calif. Publ. Zool. 50: 195–464.
This exhaustive study includes all Arizona races.

——— 1951. Generic placement of the Rufous-winged Sparrow. Wilson Bull. 63: 47–48.
Comments on writings of early observers of the species in Arizona.

Raitt, R. J.

1967. Relationships between black-eared and plain-eared forms of Bushtits *(Psaltriparus).* Auk 84: 503–528.
All Bushtits are considered to be *Psaltriparus minimus;* Arizona black-eared males are included in *P. m. lloydi.*

Rea, A. M.

1970. Status of the Summer Tanager on the Pacific slope. Condor 72: 230–233.

Includes measurements of Arizona specimens of *Piranga rubra cooperi,* but no localities are listed.

[Ridgway, R.]

1873. Revision of the falconine genera *Micrastur, Geranospiza* and *Rupornis,* and the strigine genus *Glaucidium.* Proc. Boston Soc. Nat. Hist. 16: 73–106.

Includes description of *G. passerinum* var. *californicum,* female from Prescott, Arizona. Range of *G. ferrugineum* includes Arizona.

Ridgway, R.

1873. On some new forms of American birds. Amer. Nat. 7: 602–619.

Describes *Catherpes Mexicanus* var. *conspersus* and *Peucaea aestivalis* var. *Arizonae.*

———

1882. Critical remarks on the Tree-creepers *(Certhia)* of Europe and North America. Proc. U.S. Nat. Mus. 5: 111–116.

Describes *C. familiaris montana;* range includes Arizona.

———

1882. Descriptions of some new North American birds. Proc. U.S. Nat. Mus. 5: 343–346.

Includes *Lophophanes inornatus griseus;* range in Arizona.

———

1884. Descriptions of some new North American birds. Proc. Biol. Soc. Wash. 2: 89–95.

Includes first descriptions of *Myiarchus mexicanus magister* and *Myiarchus lawrencei olivascens.*

———

1884. Remarks on the type specimens of *Muscicapa fulvifrons,* Giraud, and *Mitrephorus pallescens,* Coues. Proc. Biol. Soc. Wash. 2: 108–110.

———

1885. *Icterus cucullatus,* Swainson, and its geographical variations. Proc. U.S. Nat. Mus. 8: 18–19.

The Arizona form is named *I. c. nelsoni;* type from Tucson.

———

1885. Description of a new Cardinal Grosbeak from Arizona. Auk 2: 343–345.

Cardinalis cardinalis superbus from Fuller's Ranch, near Tucson.

———

1885. On *Junco cinereus* (Swains.) and its geographical races. Auk 2: 363–364.

Describes *J. c. palliatus,* type from Mt. Graham, Arizona.

———

1885. Some emended names of North American birds. Proc. U.S. Nat. Mus. 8: 354–356.

Concerns some Arizona species.

Ridgway, R.
1886. Arizona Quail. Forest and Stream 25: 484.
Concerns *Colinus ridgwayi* and *Ortyx graysoni;* not identical.

———
1887. Description of two new races of *Pyrrhuloxia sinuata* Bonap. Auk 4: 347.
Describes *P. s. beckhami;* range includes southern Arizona.

———
1887. Description of a new *Psaltriparus* from southern Arizona. Proc. U.S. Nat. Mus. 10: 697.
P. santaritae, type from the Santa Rita Mountains, Arizona.

———
1888. Description of a new western subspecies of *Accipiter velox* (Wils.) and subspecific diagnosis of *A. cooperi mexicanus* (Swains.). Proc. U.S. Nat. Mus. 11: 92.
A. v. rufilatus, "western North America . . ." and *A. c. mexicanus,* "western United States."

———
1894. On geographical variation in *Sialia mexicana* Swainson. Auk 11: 145–160.
Contains descriptions of Arizona specimens of *S. m. bairdi.*

———
1895. On the correct subspecific names of the Texan and Mexican Screech Owls. Auk 12: 389–390.
The name *Megascops asio cineraceus* is proposed.

———
1898. Descriptions of supposed new genera, species, and subspecies of American birds. I. Fringillidae. Auk 15: 223–230.
Describes *Amphispiza bilineata deserticola,* type from Tucson, Arizona.

———
1898. Description of a new species of Hummingbird from Arizona. Auk 15: 325–326.
Atthis morcomi, type from the Huachuca Mountains, Arizona.

———
1901. New birds of the families Tanagridae and Icteridae. Proc. Wash. Acad. Sci. 3: 149–155.
Describes *Agelaius phoeniceus fortis.*

———
1903. Descriptions of new genera, species and subspecies of American birds. Proc. Biol. Soc. Wash. 16: 105–111.
Includes *Vireo bellii arizonae,* type from Tucson, Arizona.

———
1906. "*Atratus* versus *Megalonyx.*" Condor 8: 53.
Pipilo maculatus montanus Swarth is considered a synonym of *P. m. megalonyx* Baird.

———
1906. "*Atratus* versus *Megalonyx.*" Condor 8: 100.
Regards *Pipilo maculatus montanus* as valid.

———
1911. Diagnoses of some new forms of Picidae. Proc. Biol. Soc. Wash. 24: 31–35.
Colaptes chrysoides mearnsi, type from Quitovaquito, Arizona.

———
1915. Descriptions of some new forms of American Cuckoos, Parrots, and Pigeons. Proc. Biol. Soc. Wash. 28: 105–107.
Includes *Melopelia asiatica mearnsi,* type from five miles north of Nogales, Arizona.

Rogers, C. H.
1939. A new Swift from the United States. Auk 56: 465–468.
Measurements of Arizona specimens of *A. s. saxatilis* listed.

Rylander, M. K. and E. G. Bolen
1970. Ecological and anatomical adaptations of North American Tree Ducks. Auk 87: 72–90.
No subspecies in the Fulvous Tree Duck are recognized and only two in the Black-bellied Tree Duck: *Dendrocygna autumnalis autumnalis* and *D. a. discolor.*

Schiebel, G.
1906. Die phylogenese der *Lanius*-Arten. Jour. für Ornith. 54: 1–77.
Contains only a brief discussion of the forms of *L. ludovicianus.*

Schorger, A. W.
1970. A new subspecies of *Meleagris gallopavo.* Auk 87: 168–170.
M. g. tularosa from northeastern Arizona and Tularosa Cave, near Reserve, New Mexico.

Scott, W. E. D.
1885. Early spring notes from the mountains of southern Arizona. Auk 2: 348–356.
Accounts of 36 species and description of *Troglodytes aëdon marianae,* new subspecies from the Santa Catalina Mountains.

———
1885. A mule bird. Forest and Stream 23: 484.
A hybrid between the Red-shafted and Gilded Flickers collected by H. Brown near Tucson.

Selander, R. K.
1964. Speciation in Wrens of the genus *Campylorhynchus.* Univ. Calif. Publ. Zool. 74: 1–259.
The name *Campylorhynchus brunneicapillus anthonyi* is revived for the Arizona race.

——— and R. W. Dickerman
1963. The "nondescript" Blackbird from Arizona: an intergeneric hybrid. Evolution 17: 440–448.
A hybrid between *Cassidix mexicanus* and *Agelaius phoeniceus.*

——— and R. F. Johnston
1967. Evolution in the House Sparrow. I. Intrapopulation variation in North America. Condor 69: 217–258.
Many Arizona specimens were used in the study.

Sennett, G. B.
1888. Notes on the *Peucaea ruficeps* group, with description of a new subspecies. Auk 5: 40–42.
Describes *P. r. scottii,* type from Pinal County, Arizona.

Short, L. L., Jr.
1965. Hybridization in the Flickers *(Colaptes)* of North America. Bull. Amer. Mus. Nat. Hist. 129: 307–428.
North American Flickers are considered conspecific. Includes important data on Arizona distribution and hybridization.

———
1968. Variation of Ladder-backed Woodpeckers in southwestern North America. Proc. Biol. Soc. Wash. 81: 1–10.
Mohavensis and *yumanensis* are regarded as synonyms of *cactophilus.*

Short, L. L., [Jr.] and J. J. Morony, Jr.
1970. A second hybrid Williamson's x Red-naped Sapsucker and an evolutionary history of Sapsuckers. Condor 72: 310–315.
Oberholser's (1930) hybrid is fully described, and also a variant female Williamson's Sapsucker.

Short, L. L., Jr. and A. R. Phillips
1966. More hybrid Hummingbirds from the United States. Auk 83: 253–265.
Eugenes fulgens x *Cynanthus latirostris* from the Huachuca Mountains, Arizona.

[Sizer, J. W.]
1966. [Hybrid Quail.] Wildlife Views 13 (6): 1.
Illustration of hybrid of Gambel and Scaled Quail on front cover.

———
1967. [Comment on hybrid Quail.] Wildlife Views 14(2): 23.
Hybrids are reported each year in the area where Gambel and Scaled Quail ranges overlap.

Snyder, L. L. and H. G. Lumsden
1951. Variation in *Anas cyanoptera.* Occ. Papers, Roy. Ont. Mus. Zool. 10: 1–18.
The Arizona birds are *A. c. septentrionalium.*

Stephens, F.
1895. Descriptions of two new subspecies of California birds. Auk 12: 371–372.
Callipepla gambeli deserticola ranges "east through the Colorado Valley."

Storer, R. W. and D. A. Zimmerman
1959. Variation in the Blue Grosbeak *(Guiraca caerulea)* with special reference to the Mexican populations. Occas. Papers Mus. Zool. Univ. Mich. 609: 1–13.
Critical review; includes measurements of Arizona specimens.

Sutton, G. M.
1934. A new Bewick's Wren from the western panhandle of Oklahoma. Auk 51: 217–220.
Thryomanes bewicki niceae is believed to breed in the mountains of northern Arizona.

——— and A. R. Phillips

1942. The northern races of *Piranga flava.* Condor 44: 277–279.

Arizona birds are *P. f. hepatica;* the races *P. f. oreophasma* and *P. f. zimmeri* are not recognized.

Swarth, H. S.

1905. *Atratus* versus *Megalonyx.* Condor 7: 171–174.

First description of *Pipilo maculatus montanus,* type from Huachuca Mountains.

———

1910. Two new Owls from Arizona with description of the juvenal plumage of *Strix occidentalis occidentalis* (Xantus). Univ. Calif. Publ. Zool. 7: 1–8.

Describes *Otus asio gilmani* and *Strix occidentalis huachucae.*

———

1912. Report on a collection of birds and mammals from Vancouver Island. Univ. Calif. Publ. Zool. 10: 1–124.

Critical remarks on *Geothlypis trichas scirpicola* and *Chordeiles virginianus henryi* of Arizona.

———

1913. The status of Lloyd's Bush-tit as a bird of Arizona. Auk 30: 399–401.

Records of *Psaltriparus melanotis lloydi* in Arizona are shown to pertain to the juveniles of *P. plumbeus.*

———

1913. A revision of the California forms of *Pipilo maculatus* Swainson, with description of a new subspecies. Condor 15: 167–175.

Specimen from Fort Yuma is *P. m. curtatus.*

———

1914. The California forms of the genus *Psaltriparus.* Auk 31: 499–526.

Describes variation in *P. plumbeus* from Arizona.

———

1915. The status of the Arizona Spotted Owl. Condor 17: 15–19.

Strix occidentalis huachucae is considered a valid race; four Arizona specimens are listed.

———

1916. The Sahuaro Screech Owl as a recognizable race. Condor 18: 163–165.

Otus asio gilmani is a valid race restricted to the Lower Sonoran zone.

———

1917. Geographical variation in *Sphyrapicus thyroideus.* Condor 19: 62–65.

The Rocky Mountain subspecies, including birds from Arizona, is regarded as *S. t. nataliae.*

Taylor, W. K.

1970. Some taxonomic comments on the genus *Auriparus.* Auk 87: 363–366.

Auriparus is related to *Coereba.* Notes on behavior in Arizona.

Todd, W. E. C.

1916. Preliminary diagnoses of fifteen apparently new neotropical birds. Proc. Biol. Soc. Wash. 29: 95–98.

Includes description of *Falco fusco-coerulescens septentrionalis,* type from Fort Huachuca, Arizona.

Traylor, M. A., Jr.
1949. Notes on some Vera Cruz birds. Fieldiana 31: 269–275.
Parus sclateri eidos of Chiricahua Mountains not distinguishable from Vera Cruz specimens.

Twomey, A. C.
1947. Critical notes on some western Song Sparrows. Condor 49: 127–128.
M. m. bendirei is regarded as a good race; breeding range of *M. m. fallax* in Arizona is discussed.

van Rossem, A. J.
1927. The Arizona race of the Sulphur-bellied Flycatcher. Condor 29: 126.
First description of *Myiodynastes luteiventris swarthi,* type from the Huachuca Mountains, Arizona.

———

1929. The races of *Sitta pygmaea* Vigors. Proc. Biol. Soc. Wash. 42: 175–178.
Description of *S. p. melanotis,* type from the Chiricahua Mountains, Arizona.

———

1929. The status of some Pacific Coast Clapper Rails. Condor 31: 213–215.
Rallus yumanensis is regarded as a race of *R. obsoletus;* range along the lower Colorado River.

———

1930. The races of *Auriparus flaviceps* (Sundevall). Trans. San Diego Soc. Nat. Hist. 6: 199–202.
A. f. flaviceps and *A. f. ornatus:* ranges include southern Arizona.

———

1930. The Sonora races of *Camptostoma* and *Platypsaris.* Proc. Biol. Soc. Wash. 43: 129–132.
Camptostoma imberbe ridgwayi is regarded as valid; *Platypsaris aglaiae richmondi* described as new; ranges include Arizona.

———

1930. Critical notes on some Yellowthroats of the Pacific southwest. Condor 32: 297–300.
Describes *G. t. chryseola,* new subspecies ranging into southeastern Arizona.

———

1930. A northwestern race of the Mexican Goshawk. Condor 32: 303–304.
Asturina plagiata maxima, new subspecies from southern Arizona.

———

1931. Concerning some western races of *Polioptila melanura.* Condor 33: 35–36.
The Arizona form is *P. m. lucida.*

———

1934. Notes on some types of North American birds. Trans. San Diego Soc. Nat. Hist. 7: 347–362.
New subspecies: *Pyrocephalus rubinus flammeus,* and *Tangavius aeneus milleri,* ranging into Arizona.

———

1934. A northwestern race of the Varied Bunting. Trans. San Diego Soc. Nat. Hist. 7: 369–370.
Description of *Passerina versicolor dickeyae;* range includes southern Arizona.

———

1935. Notes on the forms of *Spizella atrogularis*. Condor 37: 282–284.
The Arizona race is *S. a. evura.*

———

1936. Remarks stimulated by Brodkorb's "Two new subspecies of the Red-shafted Flicker." Condor 38: 40.
Brodkorb's *C. c. canescens* and *C. c. chihuahuae* are regarded as intermediate forms.

———

1937. The Ferruginous Pigmy Owl of northwestern Mexico and Arizona. Proc. Biol. Soc. Wash. 50: 27–28.
Describes *Glaucidium gnoma cactorum;* range includes southern Arizona.

———

1938. A Colorado Desert race of the Summer Tanager. Trans. San Diego Soc. Nat. Hist. 9: 13–14.
First description of *Piranga rubra hueyi;* range includes southwestern Arizona.

———

1938. A Mexican race of the Goshawk (*Accipiter gentilis* [Linnaeus]). Proc. Biol. Soc. Wash. 51: 99.
Describes *A. g. apache,* type from Paradise, Cochise County, Arizona.

———

1939. XXXII. Some new races of birds from Mexico. Annals and Mag. Nat. Hist. 11th ser. 4: 439–443.
First description of *Anhinga a. minima,* the northern race, and *Polyborus cheriway ammophilus,* occurring into southern Arizona.

———

1939. A race of the Rivoli Humming Bird from Arizona and northwestern Mexico. Proc. Biol. Soc. Wash. 52: 7–8.
First description of *Eugenes fulgens aureoviridis.*

———

1940. Notes on some North American birds of the genera *Myiodynastes, Pitangus,* and *Myiochanes.* Trans. San Diego Soc. Nat. Hist. 9: 79–86.
Describes *Myiochanes virens placens,* type from Madera Canyon, Santa Rita Mountains, Arizona.

———

1941. A race of the Poor-will from Sonora. Condor 43: 247.
Phalaenoptilus nuttallii adustus, new subspecies, extends into "extreme southern Arizona."

———

1942. A western race of the Tooth-billed Tanager. Auk 59: 87–89.
Piranga flava zimmeri, new name for the Arizona race.

van Rossem, A. J.

1942. Four new Woodpeckers from the western United States and Mexico. Condor 44: 22–26.

Centurus uropygialis albescens and *Dryobates scalaris yumanensis* are new subspecies from the Colorado River valley.

——

1945. A distributional survey of the birds of Sonora, Mexico. Occas. Papers Mus. Zool. Louisiana State Univ. 21: 1–379.

Includes references to species along the Arizona-Sonora border; describes *Icterus bullockii parvus;* range includes lower Colorado River valley.

——

1946. Two new races of birds from the lower Colorado River valley. Condor 48: 80–82.

Toxostoma dorsale coloradense, Pipilo aberti dumeticolus, new subspecies from Yuma.

——

1946. Two new races of birds from the Harquahala Mountains, Arizona. Auk 63: 560–563.

Pipilo fuscus relictus and *Aimophila ruficeps rupicola* described as new.

——

1947. The distribution of the Yuma Horned Lark in Arizona. Condor 49: 38–40.

Includes map of distribution of *O. a. leucansiptila, occidentalis,* and *adusta* in Arizona.

——

1947. Two races of the Bridled Titmouse. Fieldiana 31: 87–92.

First description of *Parus wollweberi phillipsi,* type from Pajaritos Mountains, Arizona.

—— and the Marquess Hachisuka

1937. A northwestern race of the Mexican Black Hawk. Trans. San Diego Soc. Nat. Hist. 8: 361–362.

Description of *Buteogallus anthracinus micronyx;* range includes Arizona.

—— and the Marquess Hachisuka

1938. A new race of Cliff Swallow from northwestern Mexico. Trans. San Diego Soc. Nat. Hist. 9: 5–6.

Description of *Petrochelidon albifrons minima;* range includes southern Arizona.

Van Tyne, J.

1925. An undescribed race of Phainopepla. Occas. Papers Boston Soc. Nat. Hist. 5: 149–150.

Phainopepla nitens lepida; range includes Arizona.

—— and G. M. Sutton

1937. The birds of Brewster County, Texas. Univ. Mich. Mus. Zool. Misc. Publ. 37: 1–119.

Figures an Arizona specimen of *Buteo jamaicensis calurus;* compares *Aphelocoma sordida arizonae* with *A. s. couchii.*

Webster, J. D.
1957. A new race of Wood Pewee from Mexico. Proc. Indiana Acad. Sci. 66: 337–340.
Specimens of *Contopus sordidulus veliei* listed from Arizona.

———
1958. Systematic notes on the Olive Warbler. Auk 75: 469–473.
Map of range of *P. t. arizonae;* specimens from Arizona examined.

———
1959. A revision of the Botteri Sparrow. Condor 61: 136–146.
Includes synopsis of *A. b. arizonae* with list of specimens from Arizona.

———
1961. A revision of Grace's Warbler. Auk 78: 554–566.
Discussion and diagnosis of all the races, including *D. g. graciae* of Arizona.

———
1962. Systematic and ecologic notes on the Olive Warbler. Wilson Bull. 74: 417–425.
Arizona specimens discussed.

———
1963. A revision of the Rose-throated Becard. Condor 65: 383–399.
Gives synopsis of *P. a. albiventris* and list of specimens of which one is from Arizona.

Wetmore, A.
1941. Notes on birds of the Guatemalan highlands. Proc. U.S. Nat. Mus. 89: 523–581.
Falco sparverius phalaena is considered not recognizable; all northern birds are *F. s. sparverius.*

———
1947. The races of the Violet-crowned Hummingbird, *Amazilia violiceps.* Jour. Wash. Acad. Sci. 37: 103–104.
Amazilia salvini reported from the Huachuca Mountains proves to be *Amazilia violiceps ellioti.*

———
1949. Geographical variation in the American Redstart *(Setophaga ruticilla).* Jour. Wash. Acad. Sci. 39: 137–139.
S. r. tricolora is the Arizona form.

———
1964. A revision of the American Vultures of the genus *Cathartes.* Smiths. Misc. Coll. 146(6): 1–18.
Cathartes aura meridionalis (formerly *C. a. teter*) is assigned to northern Arizona; *C. a. aura* to southern Arizona; *C. a. septentrionalis* is casual.

——— and J. L. Peters
1922. A new genus and four new subspecies of American birds. Proc. Biol. Soc. Wash. 35: 41–46.
Describes *Dendrocygna bicolor helva;* range includes Arizona.

Zimmer, J. T.

1929. A study of the Tooth-billed Red Tanager, *Piranga flava.* Field Mus. Nat. Hist. Zool. Series 17: 169–219.

Critical notes on *P. f. hepatica,* with synonymy, description and distribution. Arizona specimens listed.

———

1939. Studies of Peruvian birds, 30. Notes on the genera *Contopus, Empidonax, Terenotriccus,* and *Myiobius.* Amer. Mus. Novit. 1042: 1–13.

Contopus pertinax is considered conspecific with *C. fumigatus;* specimens from Arizona examined.

2. Distribution and Migration

General Works

This section contains the larger general ornithological publications, offering in some degree information on taxonomy, diagnoses, distribution, and migration of Arizona birds. Some are monographs of particular orders of birds; others are regional treatises of wide areas, such as the western United States, North and South America, and the world. A few are popular identification manuals.

American Ornithologists' Union Committee

1886. Check-list of North American birds and code of nomenclature. Amer. Ornith. Union. 392 pp. [1st edition.]; 1895. 2nd and revised edition, Amer. Ornith. Union. New York. 372 pp.; 1910. 3rd edition (rev.), Amer. Ornith. Union. New York. 430 pp.; 1931. 4th edition. Lancaster, Pa. 526 pp.; 1957. 5th edition. Baltimore. 691 pp.

[American Ornithologists' Union] F. M. Chapman

1888. List of additions to the North American avifauna and of eliminations and changes in nomenclature proposed since the publication of the A. O. U. Check-list. Auk 5: 393–402.

American Ornithologists' Union Committee

1890. Second supplement to the American Ornithologists' Union Check-list of North American birds. Auk 7: 60–66; 1891. 3rd suppl. Auk 8: 83–90; 1892. 4th suppl. Auk 9: 105–108; 1893. 5th suppl. Auk 10: 59–63; 1894. 6th suppl. Auk 11: 46–51; 1895. 7th suppl. Auk 12: 163–169; 1897. 8th suppl. Auk 14: 117–135; 1899. 9th suppl. Auk 16: 97–133; 1901. 10th suppl. Auk 18: 295–320; 1902. 11th suppl. Auk 19: 315–342; 1903. 12th suppl. Auk 20: 331–368; 1904. 13th suppl. Auk 21: 411–424; 1908. 14th suppl. Auk 25: 343–399; 1909. 15th suppl. Auk 26: 294–303; 1912. 16th suppl. Auk 29: 380–386; 1920. 17th suppl. Auk 37: 439–449; 1923. 18th suppl. Auk 40: 513–525; 1944. 19th suppl. Auk 61: 441–464; 1945. 20th suppl. Auk 62: 436–449; 1946. 21st suppl. Auk 63: 428–432; 1947. 22nd suppl. Auk 64: 445–452; 1948. 23rd suppl. Auk 65: 438–443; 1949. 24th suppl. Auk 66: 281–285; 1950. 25th suppl. Auk 67: 368–370; 1951. 26th suppl. Auk 68: 367–369; 1952. 27th suppl. Auk 69: 308–312; 1953. 28th suppl. Auk 70: 359–361; 1954. 29th suppl. Auk 71: 310–312; 1955. 30th suppl. Auk 72: 292–295; 1956. 31st suppl. Auk 73: 447–449.

Bailey, F. M.
1917. Handbook of birds of the western United States. 5th ed. revised. Houghton Mifflin Co. 574 pp. (1st edition, 1902.)

Baird, S. F.
1864–1872. Review of American birds in the museum of the Smithsonian Institution. Part I, North and Middle America. Smiths. Misc. Coll. 181: 1–478.
Includes the original description of *Dendroica graciae,* type from Fort Whipple, Arizona.

———, T. M. Brewer and R. Ridgway
1874. A history of North American birds. Land birds. Vol. 1, 596 pp., vol. 2, 590 pp., vol. 3, 560 pp.

———, T. M. Brewer and R. Ridgway
1884. The water birds of North America. (Mem. Mus. Comp. Zool. Harvard, vol. 12). Little, Brown and Co., Boston. 2 vols., 1–537 and 1–552.
Mentions Trumpeter Swan at Fort Mohave in vol. 1, p. 431.

———, J. Cassin and G. N. Lawrence
1858. Birds. In Pacific Railroad Reports, Explorations and Surveys for a railroad route from the Mississippi River to the Pacific Ocean. War Dept., Washington, D.C., 9: 1–1005.

Brandt, H.
1951. Arizona and its bird life. The Bird Research Foundation, Cleveland. 1–725.
Extensive observations, chiefly of nesting, of southeastern Arizona birds.

Brown, L. and D. Amadon
1968. Eagles, Hawks and Falcons of the world. McGraw-Hill Book Co. 2 vols. 945 pp.

Chapman, F. M.
1907. The Warblers of North America. D. Appleton and Co., New York. 1–306.
Includes migration of Arizona species.

Cooke, W. W.
1904. Distribution and migration of North American Warblers. U.S. Dept. Agric., Div. Biol. Surv. Bull. 18: 1–142.
Contains numerous Arizona dates; first record of Prothonotary Warbler.

———
1906. Distribution and migration of North American Ducks, Geese, and Swans. U.S. Dept. Agric. Biol. Surv. Bull. 26: 1–90.

Cory, C. B.
1918. Catalogue of birds of the Americas. Field Mus. Nat. Hist. Zool. Series, vol. 13, part 2, no. 1: 1–315.
Includes Arizona species in Strigiformes, Psittaciformes, Coraciiformes, Caprimulgidae, Cypselidae and Trochilidae. 1919. Vol. 13, part 2, no. 2: 314–

607, includes Arizona species in Trogonidae, Cuculidae, and Picidae.
Note: Thirteen of the 15 parts in the "Catalogue of birds of the Americas," 1918–1949, contain references to Arizona birds. For continuation see Cory and Hellmayr, 1927; Hellmayr, 1929, 1934, 1935, 1936, 1937, 1938; Hellmayr and Conover, 1942, 1948a, 1948b, 1949.

——— and C. E. Hellmayr
1927. Catalogue of birds of the Americas. Field Mus. Nat. Hist. Zool. Series, vol. 13, part 5: 1–517.
Includes Arizona species in Tyrannidae.

Coues, E.
1872. Key to North American birds. Dodd and Mead, New York: 1–361; 1884. 2nd edition; 1887. 3rd edition; 1890. 4th edition; 1903. 5th edition, rev., 2 vols.

———
1874. Birds of the northwest: a hand-book of the ornithology of the region drained by the Missouri River and its tributaries. U.S. Geol. Surv. of the Terr. Washington., Misc. Publ. 3: 1–791.

———
1878. Birds of the Colorado Valley. A repository of scientific and popular information concerning North American ornithology. U.S. Geol. Surv. of the Terr. Washington., Misc. Publ. 11: 1–807.

Crispens, C. G., Jr.
1960. Quails and Partridges of North America. A bibliography. U. Wash. Press, Pub. Biol. 20: 1–125.
Includes all Arizona species.

Dawson, W. L.
1923. The birds of California. Booklovers' ed. South Moulton Co. San Diego, Los Angeles, San Francisco. 4 vols. 2121 pp.
Vol. II contains photographs of nests of Verdin (p. 627), Palmer's Thrasher (p. 696), with observations, Vermilion Flycatcher (p. 912), all "taken in Arizona" and observations on the Lead-colored Bush-tit in the Huachuca Mountains, Arizona (pp. 637–638). Vol. III contains photographs of Elf Owl habitat (p. 1134), and White-winged Doves in a tree (p. 1167), "taken in Arizona."

Deignan, H. G.
1961. Type specimens of birds in the United States National Museum. U.S. Nat. Mus. Bull. 221: 1–718.

Delacour, J.
1954. The waterfowl of the world. Country Life Limited. London. Vol. I: 1–284; 1956, vol II: 1–232; 1959, vol. III: 1–270.
Includes maps of distribution.

Fisher, A. K.
1893. Report on the ornithology of the Death Valley expedition of 1891, comprising notes on the birds observed in southern California, southern Nevada, and parts of Arizona and Utah. U.S. Dept. Agric. Div. Ornith. and Mamm. In North American Fauna 7: 7–158.

Friedmann, H.
1950. The birds of North and Middle America. U.S. Nat. Mus. Bull. 50 (11): 1–793.
Contains the Falconiformes.

———, L. Griscom, and R. T. Moore
1950. Distributional check-list of the birds of Mexico. Part I. Pac. Coast Avi. 29: 1–202. Note: For Part II, see A. H. Miller, 1957, Sec. 2.
Mentions Arizona distribution in many of the general accounts of species.

Gadow, H.
1883. Catalogue of the Passeriformes or perching birds in the collection of the British Museum. Cichlomorphae: Pt. V: Paridae, Laniidae, and Certhiomorphae. Vol. 8: 1–385.

Goldman, E. A. and R. T. Moore
1945. The biotic provinces of Mexico. Jour. Mamm. 26: 347–360.
Two of the provinces extend into Arizona.

Goodwin, D.
1967. Pigeons and Doves of the world. Trustees of the British Museum (Natural History), London. Publ. no. 663: 1–446.
A compilation.

Gray, A. P.
1958. Bird hybrids. A check-list with bibliography. Tech. Comm. no. 13, Commonwealth Bureau of Animal Breeding and Genetics, Edinburgh. Commonwealth Agric. Bur. Farnham Royal, Bucks, England: 1–390.
Localities seldom given: must be determined from original publication.

Grinnell, J., H. C. Bryant, and T. I. Storer
1918. The game birds of California. Univ. Calif. Press, Berkeley. 642 pp.
Includes brief notes on distribution along the Colorado River.

——— and A. H. Miller
1944. The distribution of the birds of California. Pac. Coast Avi. 27: 1–608.
Includes ranges of species that extend along the Colorado River.

Grossman, M. L. and J. Hamlet
1964. Birds of prey of the world. Clarkson N. Potter, Inc. New York. 496 pp.
Includes small distributional maps.

Hardy, R. and H. G. Higgins
1940. An annotated check-list of the birds of Washington County, Utah. Proc. Utah Acad. Sci. Arts and Letters, 17: 95–111.
Included here because of its proximity to northwestern Arizona.

Hargitt, E.
1890. Catalogue of the Picariae in the collection of the British Museum. Vol. 18: 1–597.

Hellmayr, C. E.
1929. Catalogue of birds of the Americas. Field Mus. Nat. Hist. Zool. Series, vol. 13, part 6: 1–258.
Includes synonymy and range of *Platypsaris aglaiae albiventris* in Arizona. 1934. vol. 13, part 7: 1–531, includes species in Corvidae, Paridae, Sittidae,

Certhiidae, Cinclidae, Troglodytidae, Mimidae, Turdidae, and Sylviidae. 1935. vol. 13, part 8: 1–541, includes Arizona species in Alaudidae, Hirundinidae, Motacillidae, Bombycillidae, Ptilogonatidae, Vireonidae, Laniidae, Sturnidae, and Compsothlypidae. 1936. vol. 13, part 9: 1–458, includes Arizona species in Thraupidae. 1937. vol. 13, part 10: 1–228, includes Arizona species in Icteridae. 1938. vol. 13, part 11: 1–662, includes Arizona species in Ploceidae and Fringillidae.

——— and B. Conover
1942. Catalogue of birds of the Americas. Field Mus. Nat. Hist. Zool. Series, vol. 13, part 1, no. 1: 1–636, includes Arizona species in Tetraonidae, Phasianidae, Meleagrididae, Gruidae, Rallidae, and Columbidae. 1948a. vol. 13, part 1, no. 2: 1–434, includes Arizona species in Gaviidae, Colymbidae, Pelecaniformes, Ciconiiformes, and Anatidae. 1948b. vol. 13, part 1, no. 3: 1–383, includes Arizona species in Charadriiformes. 1949. vol. 13, part 1, no. 4: 1–358, includes Arizona species in Falconiformes.

Henshaw, H. W.
1875. Notes upon the ornithology of the regions traversed, pp. 149–150. In App. I 1 of App. LL of Annual Report Geog. Exp. and Surv. W. 100th Merid., etc. (Wheeler).
Includes brief notes on Arizona itinerary.

———
1875. Report upon the ornithological collections made in portions of Nevada, Utah, California, Colorado, New Mexico, and Arizona, during the years 1871, 1872, 1873, and 1874. In Rep. Geog. and Geol. Surv. West 100th Merid. by G. M. Wheeler, 5 (3): 1–20, 131–507, 977–989.

Hornaday, W. T.
1908. Campfires on desert and lava. Charles Scribner's Sons, New York. 366 pp.
An account of a trip from Tucson to Pinacate Peak in November, 1907. Includes a chapter on the birds observed.

Kortright, F. H.
1943. The Ducks, Geese and Swans of North America. American Wildlife Institute, Washington. 476 pp.

Lane, J. A.
1965. A birdwatcher's guide to southeastern Arizona. L. and P. Photography, Santa Ana, Calif. 46 pp.
A good guide to important localities.

May, J. B.
1935. Hawks of North America. Nat. Assoc. Aud. Soc. 140 pp.
Includes Arizona species. Good colored illustrations with descriptions, ranges, and notes on habits.

Mayr, E. and J. C. Greenway, Jr., eds.
1960. Check-list of birds of the world. Mus. Comp. Zool. Cambridge. Vol. 9: 1–506, includes Arizona species in Alaudidae, Hirundinidae, Motacillidae, Laniidae, Bombycillidae, Cinclidae, Mimidae, and Troglodytidae. 1962. vol. 15: 1–315, includes Arizona species in Ploceidae, Sturnidae and Corvidae.

Mayr, E. and R. A. Paynter, Jr., eds.
1964. Check-list of birds of the world. Mus. Comp. Zool. Cambridge. Vol. 10: 1–502.
Includes Arizona species in Turdinae and Polioptilinae.

Meinertzhagen, R.
1926. Introduction to a review of the genus *Corvus*. Novit. Zool. 33: 57–121.
Synonymy, description, and distribution of all Arizona species.

Miller, A. H., ed. in chief.
1957. Distributional check-list of the birds of Mexico. Part II. Pac. Coast Avi. 33: 1–436.
Mentions Arizona distribution in many of the general accounts.

Ogilvie-Grant, W. R.
1893. Catalogue of the game birds (Pterocletes, Gallinae, Opisthocomi, Hemipodii) in the collection of the British Museum. Vol. 22: 1–585.

Palmer, R. S., ed.
1962. Handbook of North American birds. Yale Univ. Press, New Haven and London. Vol. 1: 1–567.
Loons through Flamingos.

Peters, J. L.
1931. Check-list of birds of the world. Harvard Univ. Press, Cambridge. Vol. 1: 1–345, includes Arizona species in Gaviiformes, Colymbiformes, Pelecaniformes, Ciconiiformes, Anseriformes and Falconiformes. 1934. vol. 2: 1–401, includes Arizona species in Galliformes, Gruiformes and Charadriiformes. 1937. vol. 3: 1–311, includes Arizona species in Columbiformes and Psittaciformes. 1940. vol. 4: 1–291, includes Arizona species in Cuculiformes, Strigiformes, Caprimulgiformes and Apodidae. 1945. vol. 5: 1–306, includes Arizona species in Trochilidae, Trogonidae, and Alcedinidae. 1948. vol. 6: 1–259, includes Arizona species in Picidae. Note: This series continues under Mayr and Greenway, 1960, 1962 and Mayr and Paynter, 1964.

Peterson, R. T.
1941. A field guide to western birds. Houghton Mifflin Co., Boston. 240 pp. 2nd ed., 1961, 366 pp.

———
1948. Birds over America. Dodd, Mead and Co., N.Y. 342 pp. 1964, 2nd ed.
Pages 247 to 258 contain a running account of the species of birds observed in southern Arizona.

Pettingill, O. S., Jr.
1953. A guide to bird finding west of the Mississippi. Oxford Univ. Press, New York. 709 pp.
The Arizona section, pp. 3–36, is by Gale Monson.

Phillips, A., J. Marshall, and G. Monson
1964. The birds of Arizona. Univ. Ariz. Press, Tucson. 220 pp.
A thorough account of distribution and taxonomy of 423 full species of birds.

Phillips, J. C.
1923. A natural history of the Ducks. Houghton Mifflin Co., Boston and New York. Vol. 1: 1–264; vol. 2: 1–409; vol. 3: 1–383; vol. 4: 1–489.
Chiefly distribution as regards Arizona.

———
1928. Wild birds introduced or transplanted in North America. U.S. Dept. Agric. Tech. Bull. 61: 1–64.
Brief data on Arizona introductions.

——— and F. C. Lincoln
1930. American waterfowl. Their present situation and the outlook for their future. Houghton Mifflin Co., Boston and New York. 312 pp.
Range and status are given in the appendix.

Pough, R. H.
1957. Audubon western bird guide. Land, water, and game birds. Doubleday and Co., Garden City, N.Y. 316 pp.

Reed, C. A.
1904. North American birds' eggs. Doubleday, Page and Co., New York. 356 pp.

Ridgway, R.
1887. A manual of North American birds. J. B. Lippincott Co., Phila. 631 pp.

———
1892. The Hummingbirds. Separately printed from Rep. U.S. Nat. Mus. for 1890, pp. 253–383.
Descriptions and distribution of all United States species.

———
1901. The birds of North and Middle America. U.S. Nat. Mus. Bull. 50 (1): 1–715, family Fringillidae. 1902. 50 (2): 1–834, families Tanagridae, Icteridae, Coerebidae, and Mniotiltidae. 1904. 50 (3): 1–801, fifteen families, Motacillidae to Sylviidae. 1907. 50 (4): 1–973, ten families, Turdidae to Cotingidae. 1911. 50 (5): 1–859, seven families, Pteroptochidae to Trogonidae. 1914. 50 (6): 1–882, twelve families, Picidae to Bubonidae. 1916. 50 (7): 1–543, families Cuculidae, Psittacidae and Columbidae. 1919. 50 (8): 1–852, Charadriiformes. For continuation of this series see Ridgway and Friedman, 1941, 1946, and Friedmann, 1950.

——— and H. Friedman
1941. The birds of North and Middle America. U.S. Nat. Mus. Bull. 50 (9): 1–254, Gruiformes. 1946. 50 (10): 1–484, Galliformes.

Robbins, C. S., B. Bruun, and H. S. Zim
1966. Birds of North America. A guide to field identification. Golden Press, N.Y. 340 pp.

Salvadori, T.
1893. Catalogue of the Columbae, or Pigeons, in the collection of the British Museum. Vol. 21: 1–676. 1895. Vol. 27: 1–636, Chenomorphe (Palamedeae, Phoenicopteri, Anseres), Crypturi, and Ratitae.

Salvin, O. and F. D. Godman
1879–1904. Biologia Centrali-Americana. Aves, 1: 1–512.

Salvin, O. and E. Hartert
1892. Catalogue of the Picariae in the collection of the British Museum. Vol. 16: 1–703.

Saunders, H. and O. Salvin
1896. Catalogue of the Gaviae and Tubinares in the collection of the British Museum. Vol. 25: 1–475.

Sclater, P. L.
1886. Catalogue of the Passeriformes, or perching birds, in the collection of the British Museum. Fringilliformes: part II. Vol. 11: 1–431. 1888. Vol. 14: 1–494, Oligomyodae.

——— and G. E. Shelley
1891. Catalogue of the Picariae in the collection of the British Museum. Vol. 19: 1–484.

Seebohm, H. and R. B. Sharpe
1898–1902. A monograph of the Turdidae. London. Henry Sotheran and Co., Vol. 1: 1–337; vol. 2: 1–250. Vol. 1, parts IV and VI, vol. 2, parts XII and XIII include Arizona species.

Sharpe, R. B.
1875. Catalogue of the Striges or nocturnal birds of prey in the collection of the British Museum. Vol. 2: 1–326. 1877. Vol. 3: 1–343, Coliomorphe: Corvidae, Paradiseidae, Oriolidae, Dicruridae, and Prionopidae. 1885. Vol. 10, part 1: 1–682, Fringilliformes. 1888. Vol. 12, part III: 1–871, Fringilliformes. 1890. Vol. 13: 1–701, Sturniformes: Artamidae, Sturnidae, Ploceidae, Alaudidae, Atrichiidae and Menuridae. 1894. Vol. 23: 1–353, Fulicariae (Rallidae and Heliornithidae) and Alectorides (Aramidae, Eurypygidae, Mesitidae, Rhinochetidae, Gruidae, Psophiidae, and Otididae). 1896. Vol. 24: 1–794, Limicolae. Note: For continuation of this series see Gadow, 1883; Hargitt, 1890; Ogilvie-Grant, 1893; Salvadori, 1893, 1895; Salvin and Hartert, 1892; Saunders and Salvin, 1896; Sclater, 1886, 1888; Sclater and Shelley, 1891; Sharpe and Ogilvie-Grant, 1898.

———
1899–1909. A hand-list of the genera and species of birds. British Museum, London. Vol. 1: 1–303; vol. 2: 1–312; vol. 3: 1–367; vol. 4: 1–391; vol. 5: 1–694.
Accounts of distribution very brief.

——— and W. R. Ogilvie-Grant
1898. Catalogue of the Plataleae, Herodiones, Steganopodes, Pygopodes, Alcae, and Impennes in the collection of the British Museum. Vol. 26: 1–687.

——— and C. W. Wyatt
1885–94. A monograph of the Hirundinidae, or family of Swallows. Henry Sotheran and Co., London. Vol. 1: 1–354; vol. 2: 357–673.
Contains chiefly excerpts from other publications.

Shelford, V. E., ed.
1926. Naturalist's guide to the Americas. Williams and Wilkins Co., Baltimore. 761 pp.
The Arizona section, pp. 562–569, including a list of birds, is by G. A. Pearson, E. A. Goldman, F. Shreve, and C. T. Vorhies.

Simon, E.
1921. Histoire naturelle des Trochilidae. (Synopsis et Catalogue) (Encyclopédie Roret, L. Mulo, Libraire-Editeur, Paris). 416 pp.

Sprunt, A., Jr.
1955. North American birds of prey. Harper and Brothers, New York. 227 pp.
Includes brief summaries of Arizona species.

Stout, G. D., P. Mattheisson, R. V. Clem, and R. S. Palmer
1967. The Shorebirds of North America. Viking Press, New York. 270 pp.
General accounts of distribution, but little specifically Arizona.

Swann, H. K.
1922. A synopsis of the Accipitres (Diurnal Birds of Prey). 2nd ed. Wheldon and Wesley, Ltd., London. 233 pp. in 4 parts.

———
1924–1935. A monograph of the Birds of Prey. Wheldon and Wesley, Ltd., London. 14 parts; parts 6–14 edited by A. Wetmore. 487 + 448 pp.

Vaurie, C.
1959. The birds of the Palearctic fauna. A systematic reference. Order Passeriformes. London, H. F. and G. Witherby Lim. 762 pp.

Wolfe, L. R.
1938. A synopsis of North American birds of prey and their related forms in other countries. Bull. Chicago Acad. Sci. 5: 167–208.
Includes maps of distribution.

Woodhouse, S. W.
1853. Birds. In Report of an expedition down the Zuni and Colorado Rivers, by Captain L. Sitgreaves, Corps of Topographical Engineers. Washington: 58–105.

Wyman, L. E. and E. F. Burnell
1925. Field book of birds of the southwestern United States. Houghton Mifflin Co., Boston. 308 pp.

Yarrow, H. C.
1874. Progress report upon geographical and geological explorations and surveys west of the 100th meridian in 1872 . . . by G. M. Wheeler. App. E, pp. 52–55.
Brief notes on collections from the St. George, Utah, area; Cactus Wren collected.

——— and H. W. Henshaw
1874. Report upon ornithological specimens collected in the years 1871, 1872, and 1873. In Geographical and Geological Explorations and Surveys west of the 100th meridian, by G. M. Wheeler. 1874: 1–148.
Pp. 5–33, Nevada and Arizona, by Yarrow and Henshaw; pp. 34–38, Arizona and Nevada specimen list; pp. 95–148, New Mexico and eastern Arizona.

Individual Species Accounts and Annotated Lists

Papers dealing with reports of individual species occurrences at definite locations in Arizona make up the greater portion of this largest section. Annotated distributional lists from Arizona localities are listed here too.

Adams, L. A.

1907. Notes on the birds of the Baboquivari Mountains during the months of June and July. Trans. Kansas Acad. Sci. 20: 222–224.

Lists 37 species. There are many typographical errors and at least one misidentification.

Aiken, C. E. H. (Edited by E. R. Warren)

1937. Birds of the southwest. Colorado College Publ. General Series no. 212: 1–73.

Itinerary and list of birds observed on a trip from Colorado to Arizona in 1876.

Aldrich, J. W.

1943. Relationships of the Canada Jays in the northwest. Wilson Bull. 55: 217–222.

Map of distribution includes localities in Arizona where *P. c. capitalis* has been found.

———

1944. Notes on the races of the White-breasted Nuthatch. Auk 61: 592–604.

Map of range of *S. c. nelsoni* includes Arizona.

———

1946. The United States races of the Bob-white. Auk 63: 493–508.

Includes map of range of *C. v. ridgwayi;* specimens from Arizona examined.

———

1948. [Breeding distribution of the Poor-will.] Aud. Field Notes 2: 191.

Map includes localities in Arizona.

———

1949. Migration of the Lesser Scaup Duck, pp. 40–44, with 3 maps. In Migration of some North American waterfowl. U.S. Dept. Int. Fish and Wildl. Serv. Spec. Sci. Rep. (Wildl.) no. 1.

One Arizona recovery.

———

1963. Geographic orientation of American Tetraonidae. Jour. Wildl. Mgmt. 27: 528–545.

The distribution map includes *Dendragapus obscurus* in Arizona.

——— and A. J. Duvall

1955. Distribution of American gallinaceous game birds. U.S. Dept. Int. Fish and Wildl. Serv. Circ. 34: 1–30.

Includes maps of distribution of all Arizona species.

——— and A. J. Duvall
1958. Distribution and migration of races of the Mourning Dove. Condor 60: 108–128.
Includes map of distribution and a list of specimens from Arizona.

[Aldrich, J. W., et al., eds.]
1947. Distribution of North American birds. Aud. Field Notes 1: 180–182.
Map of winter distribution of the Fox Sparrow in the United States.

[Allen, A. A.]
1939. [Motion pictures of Coppery-tailed Trogons in Arizona.] [In]Minutes of Cooper Club Meetings, Condor 41: 223.

Allen, A. A.
1944. An Arizona nest of the Coppery-tailed Trogon. Auk 61: 640–642.
In the Santa Rita Mountains.

———, et al.
1944. Report of the A. O. U. committee on bird protection for 1943. Auk 61: 622–635.
Mentions 10 Roseate Spoonbills observed at the Imperial Refuge, Arizona.

Allen, J. A.
1878. Rufous-headed Sparrow (*Peucaea ruficeps*) in Texas. Bull. Nuttall Ornith. Club 3: 188–189.
Mentions range in Arizona.

———
1880. On recent additions to the ornithological fauna of North America. Bull. Nuttall Ornith. Club 5: 85–92.

———
1886. The Masked Bob-white *(Colinus ridgwayi)* in Arizona. Auk 3: 275–276.
Specimens from Arizona are discussed.

———
1886. The Masked Bob-white (*Colinus ridgwayi*) of Arizona and its allies. Bull. Amer. Mus. Nat. Hist. 1 (7), art. 16: 273–290.
Contains history, bibliography, habits, and distribution.

———
1893. List of mammals and birds collected in northeastern Sonora and northwestern Chihuahua, Mexico, on the Lumholtz archaeological expedition, 1890–92. Bull. Amer. Mus. Nat. Hist. 5 (3): 27–42.
Lists some birds from Bisbee, Arizona.

American Museum of Natural History
1957. The Southwestern Research Station of the American Museum of Natural History, Portal, Arizona. Amer. Mus. Nat. Hist.: 1–30.
Contains a list of 200 species of birds, most of them from a list compiled by A. R. Phillips.

Anderson, A. H.
1933. The Bridled Titmouse near Tucson, Arizona. Condor 35: 74.

———
1934. The Arizona state list since 1914. Condor 36: 78–83.
A summary of published additions of birds to the state list.

Anderson, A. H.
1934. Coots breeding in the Tucson region, Arizona. Condor 36: 84.

———
1934. The Turkey Vulture in southern Arizona. Wilson Bull. 46: 264.
Notes on abundance.

——— and A. Anderson
1946. The Painted Redstart near Tucson, Arizona. Condor 48: 248.

——— and A. Anderson
1947. Birds notes from southeastern Arizona. Condor 49: 89–90.
Accounts of 9 species; Wood Duck is new to the state list.

Anonymous
1928. A visit to an un-frequented part of the Grand Canyon. Grand Canyon Nature Notes 2 (9): 1–7.
An account of some of the birds seen on an 8-day trip into the western part of the park.

———
1928. Red-shafted Flicker nests on Canyon Rim. Grand Canyon Nature Notes 2 (12): 1.

———
1928. Sea Gull visits the Park. Grand Canyon Nature Notes 2 (12): 2.
Species not identified.

———
1928. Cassin Vireo observed. Grand Canyon Nature Notes 3 (1): 1.

———
1928. Dusky Grouse on South Rim. Grand Canyon Nature Notes 3 (2): 5.

———
1928. Kingfisher at Grand Canyon. Grand Canyon Nature Notes 3 (4): 1–2.

———
1929. [Flammulated Screech Owl, Eastern Kingbird, and Water Ouzel nests.] Grand Canyon Nature Notes 3 (9): 7.

———
1929. [Hepatic Tanager, California Cuckoo, and Song Sparrow reported.] Grand Canyon Nature Notes 3 (11): 5.

———
1929. [Report of Western Crow, Belted Kingfisher and other species.] Grand Canyon Nature Notes 4 (1): 5–6.

———
1929. [Bald Eagle.] Grand Canyon Nature Notes 4 (2): 6.

———
1930. [Band-tailed Pigeons and Prairie Falcons at Grand Canyon.] Grand Canyon Nature Notes 4 (1): 69–70.

———
1930. Winter Robins. Grand Canyon Nature Notes 4 (5): 33.
They ate raisins.

———
1930. [Spring bird notes at the Grand Canyon.] Grand Canyon Nature Notes 4 (6): 37.

———
1930. [Bird notes from the Grand Canyon.] Grand Canyon Nature Notes 4 (8): 56.
Mentions Red-breasted Nuthatch, Dusky Grouse, Barn Swallow, and Violet-green Swallow.

———
1930. July bird notes. Grand Canyon Nature Notes 4 (9): 62.
Account of 6 species.

———
1930. [Mexican Ground Dove and Golden-crowned Kinglet at Grand Canyon.] Grand Canyon Nature Notes 5 (1): 11.

———
1931. [Sage Thrasher in Grand Canyon National Park.] Grand Canyon Nature Notes 5 (9): 86.

———
1931. [*Empidonax traillii brewsteri* collected at Grand Canyon.] Grand Canyon Nature Notes 6 (1): 13.

———
1931. [Owls at the Grand Canyon.] Grand Canyon Nature Notes 6 (2): 21.
B. v. pallescens and *B. v. pacificus* are identified.

———
1932. [Evening Grosbeaks and Robins at Grand Canyon.] Grand Canyon Nature Notes 6 (4): 34.

———
1932. Errata. Grand Canyon Nature Notes 6 (3): 28.
Errors in vol. 5 (1): 11, vol. 5 (4): 37–38, vol. 5 (8): 83, concerning certain birds are corrected.

———
1933. Diary of early spring. Grand Canyon Nature Notes 8 (1): 140.
Arrivals of several birds at the Grand Canyon.

Arnold, J. R.
1941. Western Grasshopper Sparrow at Grand Canyon, Arizona. Condor 43: 292–293.

Audubon, M. R.
1906. Audubon's western journal: 1849–1850. Arthur H. Clark Co., Cleveland. 249 pp.
The Arizona account, pp. 146–165, is from Altar, Sonora, to the Pima Villages and then to the Colorado River.

Bailey, F. M.
1922. An Arizona feeding table. Auk 39: 474–481.
Birds observed during the winter of 1920–21 at the foot of the Santa Rita Mountains.

Bailey, F. M.
1923. Birds recorded from the Santa Rita Mountains in southern Arizona. Pac. Coast Avi. 15: 1–60.
Accounts of 174 species and subspecies.

———
1923. Notable migrants not seen at our Arizona bird table. Auk 40: 393–409.
Santa Rita Mountains.

———
1924. An Arizona valley bottom. Auk 41: 423–432.
Birds observed in the Santa Cruz Valley near Continental, Arizona.

———
1928. Birds of New Mexico. New Mex. Dept. Game and Fish. 807 pp.
Incidental mention of habits and occurrence of several species of birds in Arizona.

Bailey, V.
1903. The White-necked Raven. Condor 5: 87–89.
Observed at Willcox, Arizona.

———
1929. Life zones of the Grand Canyon. Grand Canyon Nature Notes 3 (9): 2–5.
Brief mention of the birds.

Baird, S. F.
1859. Notes on a collection of birds made by Mr. John Xantus, at Cape St. Lucas, Lower California, and now in the museum of the Smithsonian Institution. Proc. Acad. Nat. Sci. Phila. 11: 299–306.
Includes comparisons with Lower Colorado and Gila River species.

———
1859. Birds of the boundary. In Report on the United States and Mexican boundary survey by W. H. Emory, Washington, D.C. 20 (2): 1–32.

———
1860. List of birds collected on the Colorado Expedition. In Part V, Zoology: 1–6. In Report upon the Colorado River of the West, explored in 1857 and 1858 by Lt. J. C. Ives. Washington: Govt. Printing Office, 1861.

[Baker, J. H.]
1941. Help for the White-winged Dove. Aud. Mag. 43: 68, 71.
Remarks on its scarcity in Arizona.

Baker, S. and T. J. McCleneghan
1966. An Arizona economic and historic atlas. Univ. Ariz. 40 pp.
Includes outline map of Turkey distribution in Arizona.

Baldwin, G. C.
1944. Unusual records of birds from the Boulder Dam area, Nevada. Condor 46: 206–207.
List of 8 species, including a Bald Eagle.

———

1947. New records for the Boulder Dam area, Nevada. Condor 49: 85.
Observations from the west side of the Colorado River.

Bancroft, G.
1930. The breeding birds of central Lower California. Condor 32: 20–49.
Includes brief notes on the Cardinal near San Xavier Mission, Arizona.

Banks, R. C. and R. G. McCaskie
1964. Distribution and status of the Wied Crested Flycatcher in the lower Colorado River Valley. Condor 66: 250–251.

Barlow, J. C. and R. Johnson
1967. Current status of the Elf Owl in the southwestern United States. Southwestern Nat. 12: 331–332.
Chiefly a discussion of range extensions.

Barrows, W. B.
1899. The English Sparrow *(Passer domesticus)* in North America. U.S. Dept. Agric. Div. Econ. Ornith. and Mamm. Bull. 1: 1–405.
Present at Camp Huachuca in the summer of 1886.

Bartlett, K.
1942. Notes upon the routes of Espejo and Farfan to the mines in the sixteenth century. New Mex. Hist. Rev. 17 (1): 21–36.
Report of Parrots at Beaver Creek near Camp Verde.

Baumgartner, A.M.
1939. Distribution of the American Tree Sparrow. Wilson Bull. 51: 137–149.
Arizona is listed in its winter range.

Baumgartner, F. M.
1934. Bird mortality on the highways. Auk 51: 537–538.
Sennett's Thrasher is listed, erroneously, from Arizona.

———

1935. Correction. Auk 52: 104.
The record of Sennett's Thrasher (Auk 51: 537–538, 1934) is cancelled.

Behle, W. H.
1940. Distribution and characters of the Utah Red-wing. Wilson Bull. 52: 234–240.
Northern Arizona is included in the range; one specimen listed.

———

1943. Birds of Pine Valley Mountain region, southwestern Utah. Bull. Univ. Utah 34 (2): 1–85.
Included because of its proximity to the Arizona border; Arizona specimens compared.

———

1948. Birds observed in April along the Colorado River from Hite to Lee's Ferry. Auk 65: 303–306.
Running account of 37 species observed in 1947.

Behle, W. H.
1953. Migration of the White Pelicans of Gunnison Island, Great Salt Lake, Utah. Proc. Utah Acad. Sci. Arts and Letters 30: 32–38.
One banded June 11, 1948, recovered near Lowell, Cochise County, Arizona, May 9, 1950.

———
1960. The birds of southeastern Utah. Univ. Utah Biol. Ser. 12 (1): 1–56.
Includes some Glen Canyon records.

———
1963. Avifaunistic analysis of the Great Basin region of North America. Proc. 13th Int. Ornith. Cong. Ithaca, 17–24 June, 1962, vol. 2: 1168–1181.
Includes a brief account of the species of birds of the Colorado Desert and the Navajo country.

———
1966. Noteworthy records of Utah birds. Condor 68: 396–397.
The Inca Dove extends its range to 5 miles north of the Utah-Arizona border in the Beaver Dam Wash area.

———
1967. Migrant races of western Wood Pewee in Utah. Auk 84: 133–134.

———, J. B. Bushman, and C. M. Greenhalgh
1958. Birds of the Kanab area and adjacent high plateaus of southern Utah. Univ. Utah Biol. Ser. 11 (7): 1–92.
Included here because of its proximity to the Arizona Strip.

———, J. B. Bushman, and C. M. White
1963. Distributional data on uncommon birds in Utah and adjacent states. Wilson Bull. 75: 450–456.
Includes data on several species close to the Arizona border.

——— and H. G. Higgins
1959. The birds of Glen Canyon. Univ. Utah Anthro. Papers 40 (Glen Canyon Ser. 7): 107–133.
Includes Arizona records from Lee's Ferry to the Utah border.

——— and E. D. McKee
1943. Additional bird records for Grand Canyon National Park. Auk 60: 278–279.
Accounts of four species.

Bendire, C. E.
1876. Crissal Thrasher. Forest and Stream 7: 148.
Common in southern Arizona; nests early.

———
1882. American Long-eared Owl. Ornith. and Ool. 6 (11): 81–82.
Observed along Rillito Creek, Pima County, Arizona.

———
1882. Mexican Goshawk. Ornith. and Ool. 6 (11): 87–88.
Nesting near Tucson.

———
1882. Whitney Owl. Ornith. and Ool. 6 (12): 94–96.
Description of specimens from Camp Lowell, Arizona.

———

1882. The Spotted Owl. Ornith. and Ool. 7 (13): 99.
As observed near Tucson.

———

1882. The Rufous-winged Sparrow. Ornith. and Ool. 7 (16): 121–122.
An account of its habits as observed near Tucson.

———

1887. Notes on a collection of birds' nests and eggs from southern Arizona Territory. Proc. U.S. Nat. Mus. 10: 551–558.
Eight species from the vicinity of Fort Huachuca.

———

1888. Notes on the habits, nests and eggs of the genus *Glaucidium* Boie. Auk 5: 366–372.
Includes habits of *G. phalaenoides* in the Tucson region.

Bené, F.
1942. Costa Hummingbird at Papago Park, Arizona. Condor 44: 282–283.

Benson, S. B.
1934. The Black-throated Green Warbler in Arizona. Condor 36: 42.
In Toroweap Valley, Mohave County.

———

1935. A biological reconnaissance of Navajo Mountain, Utah. Univ. Calif. Publ. Zool. 40: 439–455.
Some specimens of birds were collected at the Arizona-Utah border.

Binford, L. C.
1958. First record of the Five-striped Sparrow in the United States. Auk 75: 103.
At the mouth of Madera Canyon, Santa Rita Mountains, Arizona.

Biological Survey, Bureau of
1940. The status of migratory game birds: 1939–40. U.S. Dept. Int. Bureau of Biological Survey, Wildl. Leaflet BS-165: 1–22.
Discusses the White-winged Dove in Arizona.

Bishop, L. B.
1906. *Uranomitra salvini* in Arizona. Auk 23: 337–338.
The second known example of this species collected at Palmerlee, Cochise County, Arizona.

———

1926. The distribution of the races of the Ruby-crowned Kinglet. Condor 28: 183.
Mentions Arizona specimens.

Blackford, J. L.
1940. Vertical birding. Highlights of western life zones. Bird-Lore 42: 405–413.
Includes brief notes on Arizona species.

Blake, E. R.
1942. Mexican Dipper in the Huachuca Mountains, Arizona. Auk 59: 578–579.
One collected May 28, 1903.

Bond, J.
1947. Notes on Peruvian Tyrannidae. Proc. Acad. Nat. Sci. Phila. 99: 127–154.
Mentions specimens of *Myiodynastes luteiventris vicinior* Cory from Arizona.

Bond, R. M.
1946. The Peregrine population of western North America. Condor 48: 101–116.
Includes map of distribution in western North America.

[Borden, K.]
1936. [Report of Band-tailed Pigeons in the Chiricahua Mountains, Arizona.] [In] Minutes of Cooper Club Meetings, Condor 38: 46.

Boyers, L. M.
1934. Birds collected by the 1933 Rainbow Bridge-Monument Valley Expedition. 6 pp. (No publisher listed. Berkeley?)
A list of 31 species.

Breninger, G. F.
1897. *Buteo albicaudatus* in Arizona. Auk 14: 403.
Nesting between Florence and Red Rock.

———
1898. Barn Swallows in southern Arizona. Osprey 2 (9): 117.
Nesting at Elgin, Santa Cruz County.

———
1898. The Ferruginous Pygmy Owl. Osprey 2 (10): 128.
In the Gila and Salt River region.

———
1899. White-tailed Hawk in Arizona. Auk 16: 352.
At Phoenix.

———
1901. A list of birds observed on the Pima Indian Reservation, Arizona. Condor 3: 44–46.
Eighty-six species observed during four days in September.

———
1901. The Painted Redstart. Condor 3: 147–148.
Observations in the Santa Rita and Huachuca mountains.

———
1903. Nests and eggs of *Coeligena clemenciae.* Auk 20: 435.
In the Huachuca Mountains.

———
1905. The Yellow-billed Tropic Bird near Phoenix, Arizona. Auk 22: 408.
First record of *Phaethon americanus* in Arizona.

———
1905. The English Sparrow at Tucson, Arizona. Auk 22: 417.
First record at Tucson.

Brewster, W.

1881. Notes on some birds from Arizona and New Mexico, with a description of a supposed new Whip-poor-will. Bull. Nuttall Ornith. Club 6: 65–73.

Lists 17 species, including a description of *A. v. arizonae.*

———

1881. Additions to the avi-fauna of the United States. Bull. Nuttall Ornith. Club 6: 252.

Lists *Parus meridionalis, Myiarchus cooperi,* and *M. lawrencei,* all from Arizona.

———

1882. On a collection of birds lately made by Mr. F. Stephens in Arizona. Bull. Nuttal Ornith. Club 7: 65–86; 135–147; 193–212; 1883, 8: 21–36.

———

1885. Preliminary notes on some birds obtained in Arizona by Mr. F. Stephens in 1884. Auk 2: 84–85.

Accounts of *Turdus ustulatus, Sialia sialis azurea,* and *Coeligena clemenciae.*

———

1885. Additional notes on some birds collected in Arizona and the adjoining province of Sonora, Mexico, by Mr. F. Stephens in 1884; with a description of a new species of *Ortyx.* Auk 2: 196–200.

Accounts of 19 species; first description of *Colinus ridgwayi.*

———

1887. Further notes on the Masked Bob-white *(Colinus ridgwayi).* Auk 4: 159–160.

Compares Sonora specimens.

———

1898. Occurrence of the Spotted Screech Owl *(Megascops aspersus)* in Arizona. Auk 15: 186.

In the Huachuca Mountains.

———

1902. Birds of the Cape region of Lower California. Bull. Mus. Comp. Zool. 41 (1): 1–241.

Includes comments on Arizona specimens.

Brodkorb, P.

1940. New birds from southern Mexico. Auk 57: 542–549. Mentions specimens of *Dendroica graciae graciae* from Arizona.

———

1941. The Pygmy Owl of the District of Soconusco, Chiapas. Occas. Papers Mus. Zool. Univ. Mich. 16 (450): 1–4.

Comparisons made with *Glaucidium brasilianum cactorum* of Arizona.

———

1942. Notes on some races of the Rough-winged Swallow. Condor 44: 214–217.

Discussion of *S. r. psammochrous* with specimens from Arizona.

Brodkorb, P.
1943. Geographic variation in the Band-tailed Pigeon. Condor 45: 19–20.
Measurements of specimens of *C. f. fasciata* from Arizona are given.

Brodrick, H. J.
1953. From Petrified Forest, Arizona. News from the Bird-Banders 28 (3): 28.
Banding returns of House Finch, Desert Horned Lark, and Sage Sparrow.

Brooks, A.
1935. Are small birds decreasing? Bird-Lore 37: 199–200.
Includes Arizona.

[Brown, H.]
1884. *Ortyx virginianus* in Arizona. Forest and Stream 22: 104.
A pair captured about 60 miles southwest of Tucson.

Brown, H.
1888. *Ionornis martinica* in Arizona. Auk 5: 109.
Near Tucson.

———
1899. The Scarlet Ibis *(Guara rubra)* in Arizona. Auk 16: 270.
A flock observed near Fort Lowell on September 17, 1890.

———
1899. The California Vulture in Arizona. Auk 16: 272.
Sight record at Pierce's Ferry.

———
1902. Unusual abundance of Lewis's Woodpecker near Tucson, Arizona, in 1884. Auk 19: 80–83.

———
1904. Masked Bob-white *(Colinus ridgwayi).* Auk 21: 209–213.
History, distribution, and observations on habits; believes it to be extinct in Arizona.

———
1906. The Water Turkey and Tree Ducks near Tucson, Arizona. Auk 23: 217–218.
Specimens of *Anhinga anhinga, Dendrocygna fulva,* and *D. autumnalis.*

———
1911. The English Sparrow at Tucson, Arizona. Auk 28: 486–488.
Account of its arrival and its habits.

Brown, J. L.
1963. Ecogeographic variation and introgression in an avian visual signal: the crest of the Steller's Jay *Cyanocitta stelleri.* Evolution 17: 23–39.
Includes measurements of specimens from Arizona.

Brown, P. E.
1932. Game survey of Walhalla Plateau. Grand Canyon Nature Notes 7 (4): 33–38.
Brief mention of birds observed on the snow-covered plateau.

Bruner, S. C.
1926. Notes on the birds of the Baboquivari Mountains, Arizona. Condor 28: 231–238.
Account of 103 species observed from March 18 to May 26, 1925.

Bryan, J.
1933. [Townsend Solitaire at the Grand Canyon.] Grand Canyon Nature Notes 8 (8): 216.

Bryant, H. C.
1939. Another record of the Bohemian Waxwing at Grand Canyon, Arizona. Condor 41: 123.

———
1941. A Nighthawk migration on an Arizona desert. Condor 43: 293.
On the road approaching the Grand Canyon, July 29, 1941.

———
1942. Golden Eagles visit northern Arizona desert. Condor 44: 41.

———
1945. Winter record of Red-winged Blackbirds at Grand Canyon, Arizona. Condor 47: 219.

———
1945. The status of game birds in Grand Canyon Nat'l. Park. Ariz. Wildlife and Sportsman 6 (9): 10.

———
1950. Unusual occurrence of Red-winged Blackbirds at Grand Canyon, Arizona. Condor 52: 94.

———
1952. Additions to the check-list of birds of Grand Canyon National Park, Arizona. Condor 54: 320.
Five species added to the list.

——— and A. M. Bryant
1945. Another Nighthawk migration on an Arizona desert. Condor 47: 268.
South of the Grand Canyon.

Burleigh, T. D.
1928. A brief glimpse of California bird life. Murrelet 9: 39–42.
A Shrike observed west of Flagstaff, Arizona, in December.

Campbell, B.
1932. A winter record of the Painted Redstart in Arizona. Condor 34: 192.
In Peña Blanca Canyon, Santa Cruz County.

———
1934. Bird notes from southern Arizona. Condor 36: 201–203.
Accounts of 32 species.

Campbell, I.
1934. Hermit Camp today. Grand Canyon Nature Notes 9: 277–280.
Incidental mention of several species of birds.

[Cantwell, G. G.]
1927. [Describes briefly his three months' trip in Arizona.] [In] Minutes of Cooper Club Meetings, Condor 29: 278.

Carothers, S. W.
1968. Fauna of Rio de Flag: I. Birds. Plateau 40: 101–111.
A list of 75 species; includes brief comments on the breeding species.

——— and J. R. Haldeman
1967. New records of northern Arizona birds. Plateau 40: 41–43.
First record of *Leucosticte tephrocotis*.

Carter, D. L. and R. H. Wauer
1965. Black Hawk nesting in Utah. Condor 67: 82–83.
Incidental mention of nesting at Patagonia, Arizona.

Cartwright, B. W., T. M. Shortt, and R. D. Harris
1937. Baird's Sparrow. Trans. Roy. Can. Inst. 21 (2): 153–197.
Arizona specimens listed.

[Chambers, W. L.]
1934. [Report of Black Vultures on the way from San Xavier Mission to Sells, Arizona.] [In] Minutes of Cooper Club meetings, Condor 36: 124.

[——— and H. Robertson]
1936. [Swainson and Harris Hawks observed in Arizona.] [In] Minutes of Cooper Club Meetings, Condor 38: 93.

[Chapman, F. M.]
1903. The A. O. U. trip to California. Bird-Lore 5: 99–100.
Mentions 18 species of birds observed at the Grand Canyon, Arizona.

Chapman, F. M.
1907. The new bird groups in the American Museum of Natural History. Bird-Lore 9: 168–170.
Contains a photograph of a diorama "Cactus-desert bird-life of Arizona" based on studies made at Tucson, in May 1906.

Chattin, J. E.
1968. Pacific flyway, p. 249, in J. P. Linduska ed. Waterfowl tomorrow. U.S. Dept. Int. Bur. Sport Fisheries and Wildlife, Gov. Printing Office. 784 pp.
Mentions Arizona wintering habitat.

Coale, H. K.
1894. Ornithological notes on a flying trip through Kansas, New Mexico, Arizona and Texas. Auk 11: 215–222.
Arizona places visited were Whipple Barracks, forts Verde, Mohave, Lowell, Huachuca, Grant, Thomas; and at San Carlos.

———
1915. San Lucas Verdin in Arizona. Auk 32: 106.
A. f. lamprocephalus collected at Gila Bend, April 18, 1891.

Colburn, A. E.
1917. The Goshawk in southern California and Arizona. Condor 19: 185.
Specimen from Walker, Arizona.

Cole, G. A. and M. C. Whiteside
1965. An ecological reconnaissance of Quitobaquito Spring, Arizona. Jour. Ariz. Acad. Sci. 3: 159–163.
Includes brief mention of Coots, Grebes, and Boat-tailed Grackles.

Collins, C. T.
1970. The 1969 annual banding report: a commentary. Western Bird Bander 45: 27–28.

Colton, H. S.
1930. A brief survey of the early expeditions into northern Arizona. Mus. N. Ariz. Mus. Notes 2 (9): 1–4.
The Espejo expedition found parrots at Oak Creek.

Conover, B.
1950. On *Accipiter striatus suttoni* van Rossem. Auk 67: 512.
A specimen from the Huachuca Mountains is of this race.

Cooke, M. T.
1942. Returns from banded birds. Bird-Banding 13: 110–119.
List of several species banded in Arizona.

———
1942. Returns from banded birds. Bird-Banding 13: 176–181.
List of Juncos banded in Arizona.

———
1946. Returns of banded birds: some recent records of interest. Bird-Banding 17: 63–71.
Includes several Arizona records.

———
1950. Returns from banded birds. Bird-Banding 21: 145–148.
A Cactus Wren at Tempe, Arizona.

Cooke, W. W.
1904. The migration of Warblers. Third paper. Bird-Lore 6: 57–60.
Contains migration date for Wilson Warbler in southern Arizona.

———
1904. The migration of Warblers. Fifth paper. Bird-Lore 6: 130–131.
Includes migration dates of the Grace and Black-throated Gray Warblers in Arizona.

———
1904. The migration of Warblers. Seventh paper. Bird-Lore 6: 199–200.
Contains migration dates of Hermit and Townsend Warblers in Arizona.

———
1905. The migration of Warblers. Tenth paper. Bird-Lore 7: 169–170.
Contains migration dates of Macgillivray's Warbler in Arizona.

———
1905. The migration of Warblers. Twelfth paper. Bird-Lore 7: 237–239.
Contains migration dates of the Nashville Warbler in Arizona.

———
1906. The migration of Warblers. Seventeenth paper. Bird-Lore 8: 134.
Contains migration dates of the Lucy, Virginia, and Olive Warblers in Arizona.

Cooke, W. W.

1907. The migration of Thrushes. Third paper. Bird-Lore 9: 121–125.
Contains a migration date of the Olive-backed Thrush in Arizona.

———

1907. The migration of Flycatchers. First paper. Bird-Lore 9: 264–265.
Contains migration dates of the Vermilion Flycatcher in Arizona.

———

1908. The migration of Flycatchers. Second paper. Bird-Lore 10: 16–17.
Contains migration dates of the Arkansas and Cassin Kingbirds in Arizona.

———

1908. The migration of Flycatchers. Third paper. Bird-Lore 10: 77–78.
Contains migration dates of the Hammond, Wright, Gray, Buff-breasted, and Western Flycatchers in Arizona.

———

1908. The migration of Flycatchers. Fifth paper. Bird-Lore 10: 166–170.
Contains migration dates of the Western Wood Pewee in Arizona.

———

1908. The migration of Flycatchers. Seventh paper. Bird-Lore 10: 258–259.
Contains migration dates of the Beardless and Coues' Flycatchers in Arizona.

———

1909. The migration of Flycatchers. Eighth paper. Bird-Lore 11: 12–14.
Contains migration dates of the Arizona Crested, Ash-throated, and Olivaceous Flycatchers in Arizona.

———

1909. The migration of Vireos. First paper. Bird-Lore 11: 78–82.
Contains migration dates of the Western Warbling Vireo in Arizona.

———

1909. The migration of Vireos. Second paper. Bird-Lore 11: 118–120.
Contains migration dates of the Hutton, Least, and Gray Vireos in Arizona.

———

1909. The migration of Vireos. Third paper. Bird-Lore 11: 165–168.
Contains migration dates of the Plumbeous and Cassin Vireos in Arizona.

———

1909. The migration of North American Sparrows. Bird-Lore 11: 254–260.
Contains migration dates of Brewer's Sparrow in Arizona; Tree Sparrow winters in Arizona.

———

1910. Distribution and migration of North American shorebirds. U.S. Dept. Agric. Biol. Surv. Bull. 35: 1–100.
Includes records of *Pisobia minutilla* and *Numenius americanus* in Arizona.

———

1910. The migration of North American Sparrows. Second paper. Bird-Lore 12: 12–15.
Contains migration dates of the Baird's and Western Grasshopper Sparrows in Arizona.

———

1910. The migration of North American Sparrows. Sixth paper. Bird-Lore 12: 196.
Contains migration dates of the Arkansas and Lawrence Goldfinches.

———

1910. The migration of North American Sparrows. Seventh paper. Bird-Lore 12: 240–242.
Contains migration dates of the Chestnut-collared Longspur in Arizona.

———

1911. Distribution of the American Egrets. U.S. Dept. Agric. Bur. Biol. Surv. Circ. 84: 1–5.
Range maps include Arizona.

———

1911. The migration of North American Sparrows. Eighth paper. Bird-Lore 13: 15–17.
Contains migration date of the McCown's Longspur in Arizona.

———

1911. The migration of North American Sparrows. Ninth paper. Bird-Lore 13: 83–88.
Contains migration dates of the Dickcissel and Vesper Sparrow in Arizona.

———

1911. The migration of North American Sparrows. Eleventh paper. Bird-Lore 13: 198–201.
Contains migration dates of the Blue Grosbeak in Arizona.

———

1911. The migration of North American Sparrows. Twelfth paper. Bird-Lore 13: 248–249.
Contains migration dates of Lazuli Bunting in Arizona.

———

1912. The peculiar migration of the Evening Grosbeak. Jour. Wash. Acad. Sci. 2 (1): 60–62.
Includes map of breeding localities in Arizona.

———

1912. The migration of North American Sparrows. Fifteenth paper. Bird-Lore 14: 98–105.
Contains migration dates of the Black-chinned Sparrow in Arizona.

———

1912. The migration of North American Sparrows. Sixteenth paper. Bird-Lore 14: 158–161.
Contains migration dates of the Black-headed Grosbeak in Arizona.

———

1913. The migration of North American Sparrows. Twenty-third paper. Bird-Lore 15: 236–240.
Contains migration dates of the Lincoln Sparrow in Arizona.

Cooke, W. W.

1913. Distribution and migration of North American Herons and their allies. U.S. Dept. Agric. Biol. Surv. Bull. 45: 1–70.

Includes maps of distribution and Arizona occurrences of several species.

———

1914. The migration of North American Sparrows. Twenty-seventh paper. Bird-Lore 16: 105–106.

Contains migration dates of the California Purple Finch and Cassin's Purple Finch in Arizona.

———

1914. The migration of North American Sparrows. Twenty-eighth paper. Bird-Lore 16: 176–178.

Contains distributional data on Botteri, Cassin, Rufous-winged, and Rufous-crowned Sparrows in Arizona.

———

1914. The migration of North American Sparrows. Twenty-ninth paper. Bird-Lore 16: 267–268.

Contains migration dates of the Lark Bunting in Arizona.

———

1914. The migration of North American Sparrows. Thirtieth paper. Bird-Lore 16: 351.

Contains migration dates of the Green-tailed Towhee in Arizona.

———

1914. Distribution and migration of North American Rails and their allies. U.S. Dept. Agric. Bull. 128: 1–50.

Contains data on Arizona species.

———

1914. New bird records for Arizona. Auk 31: 403–404.

Nine species added to Swarth's 1914 list of Arizona birds.

———

1915. Bird migration. U.S. Dept. Agric. Bull. 185: 1–48.

Contains migration routes of Cliff Swallow and Western Tanager through Arizona.

———

1915. Distribution and migration of North American Gulls and their allies. U.S. Dept. Agric. Bull. 292: 1–70.

Spring migration dates for Ring-billed Gull in Arizona.

———

1915. Bird migration in the Mackenzie Valley. Auk 32: 442–459.

Gives map of the migration route of the Western Tanager across Arizona, with dates.

———

1915. The migration of North American Sparrows. Thirty-second paper. Bird-Lore 17: 18–19.

Contains migration dates of the Pink-sided and Gray-headed Juncos in Arizona.

———
1915. The migration of North American birds. Bird-Lore 17: 199–203.
Contains migration date for the Blue-gray Gnatcatcher.

———
1916. The migration of North American birds. Bird-Lore 18: 14–16.
General distribution only.

Cooper, J. G.
1868. Some recent additions to the fauna of California. Proc. Calif. Acad. Sci. 4: 3–13.
Page 13: *Centrocercus urophasianus:* "Common on the eastern frontiers of California, and I have seen a fine specimen obtained as far south as the Mohave River."

———
1869. The fauna of California and its geographical distribution. Proc. Calif. Acad. Sci. 4: 61–81.
Includes a list of birds found in the Colorado Valley.

———
1869. The naturalist in California. Amer. Nat. 3: 470–481.
Observations made at Fort Mohave.

———
1870. Geological survey of California. J. D. Whitney, State Geologist. Ornithology. Vol. 1. Land birds. Edited by S. F. Baird, from the manuscript and notes of J. G. Cooper. Published by authority of the Legislature. 592 pp.
Contains observations from the vicinity of Fort Mohave, Arizona.

Coppa, J. B.
1960. Sapsuckers breeding in the Hualapai Mountains, Arizona. Condor 62: 294.

Cortopassi, A. J. and L. R. Mewaldt
1965. The circumannual distribution of White-crowned Sparrows. Bird-Banding 36: 141–169.
Contains distributional maps, also a few Arizona recoveries of banded birds.

Coues, E.
1865. Ornithology of a prairie-journey, and notes on the birds of Arizona. Ibis, 2nd ser. 1 (2): 157–165.
From Washington, D.C., to Fort Whipple, Arizona.

———
1865. [Notes on the birds observed at Fort Whipple, Arizona.] Ibis, 2nd ser. 1 (4): 535–538.
Extracts from a letter announcing additional birds observed.

———
1866. Field notes on *Lophortyx gambeli.* Ibis, 2nd ser. 2 (5): 46–55.
In "Arizona territory."

———
1866. From Arizona to the Pacific. Ibis, 2nd ser. 2 (7): 259–275.
Notes on Arizona birds observed from Fort Whipple to Fort Mohave, and down the Colorado River to Fort Yuma and return to Fort Mohave.

Coues, E.

1866. List of the birds of Fort Whipple, Arizona, with which are incorporated all other species ascertained to inhabit the Territory; with brief critical and field notes, descriptions of new species, etc. Proc. Acad. Nat. Sci. Phila.: 39–100.
Includes 244 species.

———

1868. List of birds collected in southern Arizona by Dr. E. Palmer: with remarks; Proc. Acad. Nat. Sci. Phila.: 81–85.
At Camp Grant.

———

1871. The Yellow-headed Blackbird. Amer. Nat. 5: 195–200.
At Fort Whipple; breeds in "warm parts of Arizona. . . ."

———

1871. Bullock's Oriole. Amer. Nat. 5: 678–682.
Brief mention of its arrival and departure dates in Arizona.

———

1871. The Long-crested Jay. Amer. Nat. 5: 770–775.
Includes brief notes on nesting and distribution in Arizona.

———

1872. Observations on *Picicorvus columbianus*. Ibis, 3rd ser. 2 (5): 52–59.
At Fort Whipple.

———

1872. A new bird to the United States. Amer. Nat. 6: 370.
Glaucidium ferrugineum from Tucson.

———

1872. Occurrence of Couch's Flycatcher in the United States. Amer. Nat. 6: 493.
Near Tucson.

———

1878. Note on *Passerculus bairdi* and *P. princeps*. Bull. Nuttall Ornith. Club 3: 1–3.
Mentions Arizona specimens of *P. bairdi* and gives synonymy.

———

1881. Probable occurrence of *Sarcorhamphus papa* in Arizona. Bull. Nuttall Ornith. Club 6: 248.
A sight record.

———

1892. Wintering of the Canvasback in Arizona. Auk 9: 198.
On the Verde River, near Prescott; mentions market hunting.

———

1892. Nesting of the Golden Eagle in Arizona. Auk 9: 201.
Near Prescott.

Count, E. W.

1929. The gathering of the Jay clan. Grand Canyon Nature Notes 4 (1): 2.

———

1929. Avian cliff dwellers. Grand Canyon Nature Notes 4 (3): 18.
Violet-green Swallow and White-throated Swift.

———

1930. Chisel-teeth-chatter. Grand Canyon Nature Notes 4 (10): 65–66.
Includes brief note on Long-crested Jay.

———

1930. [Hummingbirds at the Grand Canyon.] Grand Canyon Nature Notes 4 (12): 90.

Cowan, I. McT.
1938. Distribution of the races of the Williamson Sapsucker in British Columbia. Condor 40: 128–129.
Specimens of *S. t. nataliae* from Arizona are listed.

Crockett, H. L. and R. Crockett
1936. Bird records from near Phoenix, Arizona. Condor 38: 172–173.
Broad-billed Hummingbird and early nesting of a Mourning Dove.

Croft, G. Y.
1932. Cedar Waxwing *(Bombycilla cedrorum)* breeding in Utah. Auk 49: 91.
"Reports of it breeding in the mountains of Arizona . . . have been recorded." No details are given.

Crossin, R. S.
1965. The history and breeding status of the Song Sparrow near Tucson, Arizona. Auk 82: 287–288.

Crouch, J. E.
1943. Distribution and habitat relationships of the *Phainopepla.* Auk 60: 319–333.
Includes map of range in Arizona; other information is chiefly of California birds.

Daggett, F. S.
1914. Beautiful Bunting in California. Condor 16: 260.
Two collected at Blythe, California, and "fifteen or twenty" observed.

Davis, J.
1951. Distribution and variation of the Brown Towhees. Univ. Calif. Publ. Zool. 52 (1): 1–120.
Includes data of *P. fuscus* and *P. aberti* in Arizona.

———

1959. The Sierra Madrean element of the avifauna of the Cape District, Baja California. Condor 61: 75–84.
Brief references to Arizona species.

——— and L. Williams
1964. The 1961 irruption of the Clark's Nutcracker in California. Wilson Bull. 76: 10–18.
A similar irruption occurred in Arizona.

"By the Director" [W. L. Dawson]
1921. The season of 1917. Jour. Mus. Comp. Ool. 2 (1–2): 27–36.
Contains list of birds observed and eggs collected from Tucson and the Patagonia Mountains; Rufous-winged Sparrow at Indian Oasis.

Dearing, H. and M. Dearing
1946. Indigo Buntings breeding in Arizona. Condor 48: 139–140.
In Oak Creek Canyon.

Dearing, M. L.
1944. "Peck." News from the Bird-Banders 19: 45–46.
Notes on Roadrunners near Tucson.

Dickerman, R. W.
1955. Some recent Arizona bird records. Condor 57: 120–121.
Accounts of sixteen species.

——— and A. R. Phillips
1953. First United States record of *Myiarchus nuttingi.* Condor 55: 101–102.
Near Roosevelt, Gila County, Arizona.

——— and A. R. Phillips
1954. *Molothrus ater ater* in Arizona. Condor 56: 312.
At Willcox.

———, A. R. Phillips, and D. W. Warner
1967. On the Sierra Madre sparrow, *Xenospiza baileyi,* of Mexico. Auk 84: 49–60.
Includes incidental mention of *Melospiza lincolnii* from Apache County, Arizona.

Dilger, W. C.
1956. Adaptive modifications and ecological isolating mechanisms in the Thrush genera *Catharus* and *Hylocichla.* Wilson Bull. 68: 171–199.
Map of distribution of *C. guttatus* in Arizona.

———
1956. Hostile behavior and reproductive isolating mechanisms in the avian genera *Catharus* and *Hylocichla.* Auk 73: 313–353.
Map of breeding range of *C. guttatus.*

Dille, F. M.
1935. Arizona fields are virgin for bird banders. Wilson Bull. 47: 286–293.
Notes on 16 species.

———
1939. Two notable records for Arizona. Condor 41: 85.
Colaptes auratus auratus and *Chloroceryle americana septentrionalis.*

Dodge, N. N.
1940. 1939 banding at Casa Grande National Monument, Arizona. News from the Bird-Banders 15 (3): 32.

Duvall, A. J.
1942. Records from Lower California, Arizona, Idaho and Alberta. Auk 59: 317–318.
Specimens of *Dumetella carolinensis* from Springerville and the Tunitcha Mountains, Arizona.

———
1943. Breeding Savannah Sparrows of the southwestern United States. Condor 45: 237–238.
The breeding birds of the White Mountains, Arizona, belong to the race *P. s. rufofuscus.*

Dwight, J., Jr.
1900. The moult of the North American Tetraonidae (quails, partridges and grouse). Auk 17: 34–51.
Specimens of *Callipepla gambeli* and *Cyrtonyx montezumae* are mentioned.

———
1900. The moult of the North American shore birds (Limicolae). Auk 17: 368–385.
Mentions several specimens of species collected in Arizona.

———
1907. A sketch of the Thrushes of North America. Bird-Lore 9: 103–109.
Includes a map of the breeding range of *H. g. auduboni* in Arizona.

———
1918. The geographical distribution of color and of other variable characters in the genus *Junco:* a new aspect of specific and subspecific values. Bull. Amer. Mus. Nat. Hist. 38, art. 9: 269–309.
Contains a discussion of Arizona species.

Dzubin, A.
1965. A study of migrating Ross Geese in western Saskatchewan. Condor 67: 511–534.
A Ross Goose banded at Kindersley was recovered in Arizona.

Eaton, T. H., Jr. and G. Smith
1937. Birds of the Navajo Country. Project 6677-Y, Nat. Youth Administration, Berkeley, Calif. 75 pp.
Descriptions, habitat, nests, and eggs of 101 species of birds.

Eisenmann, E.
1962. Notes on Nighthawks of the genus *Chordeiles* in southern middle America, with a description of a new race of *Chordeiles minor* breeding in Panama. Amer. Mus. Novit. 2094: 1–21.
Lists specimens of *C. m. henryi* from Arizona.

Elliot, D. G.
1892. The inheritance of acquired characters. Auk 9: 77–104.
Mentions *Melospiza fasciata (-melodia) fallax* and *M. f. montana* in Arizona.

Enderson, J. H.
1965. A breeding and migration survey of the Peregrine Falcon. Wilson Bull. 77: 327–339.
Includes Arizona map of Christmas count sightings from 1947–1963.

Euler, R. C. and H. F. Dobyns
1962. Excavations west of Prescott, Arizona. Plateau 34: 69–84.
Mourning Doves observed near the site.

Evenden, F. G., Jr.
1952. Winter status of Swallows in California. Condor 54: 360–361.
Records Rough-winged Swallow along the Colorado River in February.

Evermann, B. W.
1882. Black-crested Flycatcher. Ornith. and Ool. 7: 177–179.
Phainopepla nitens in Arizona.

Evermann, B. W. and O. P. Jenkins
1888. Ornithology from a railroad train. Ornith. and Ool. 13: 65–70.
A running account of birds observed on a journey from Terre Haute, Indiana, to Guaymas, Sonora, with some remarks on Arizona species.

Fast, J. E.
1938. [Gambel Sparrow banded at Coolidge, Arizona is found at Highlands, California.] News from the Bird-Banders 13 (3): 34.

———
1939. [Banding notes from Nogales, Arizona.] News from the Bird-Banders 14 (4): 47.

Finley, W. L.
1908. Life history of the California Condor. Part 2. Historical data and range of the Condor. Condor 10: 5–10.
Herbert Brown, George F. Breninger, and O. W. Howard reported no Condors in Arizona.

——— and I. Finley
1915. Bird friends in Arizona. Bird-Lore 17: 237–245.
A running account of many of the birds observed in the Tucson area.

——— and I. Finley
1915. Birds of the cactus country. Bird-Lore 17: 334–341.
Notes on birds of the Tucson region.

[Fish and Wildlife Service.]
1942. Numbers of birds banded during the government fiscal year 1941. Bird-Banding 13: 134–141.
Mrs. Jos. T. Birchett of Tempe, Arizona, banded between 500 and 800 birds.

Fish and Wildlife Service
1946. National wildlife refuges administered by the Fish and Wildlife Service. U.S. Dept. Int. Fish and Wildl. Serv. Wildlife Leaflet 179: 1–12.
Eight Arizona refuges are listed with their important species of animals.

Fisher, A. K.
1892. *Myiarchus nuttingi* in Arizona. Auk 9: 394.
Misidentified; see Nelson, 1904. Specimens are *M. cinerascens.*

———
1893. The Hawks and Owls of the United States in their relation to agriculture. U.S. Dept. Agric. Div. Ornith. and Mamm. Bull. 3: 1–210.
Includes notes on some Arizona specimens.

———
1894. The capture of *Basilinna leucotis* in southern Arizona. Auk 11: 325–326.
In the Chiricahua Mountains.

———
1903. A partial list of the birds of Keams Canyon, Arizona. Condor 5: 33–36.
Thirty-nine species recorded in July 1894.

———

1904. [Review of Swarth's Birds of the Huachuca Mountains, Arizona.] Condor 6: 80–81.

Twelve species are added to the list, including *Dendroica virens,* first record for Arizona.

Fowler, F. H.

1903. Stray notes from southern Arizona. Condor 5: 68–71.

Habits and occurrence of *Cyrtonyx montezumae, Columba fasciata, Trogon ambiguus,* and *Urubitinga anthracina.*

———

1903. Stray notes from southern Arizona. Condor 5: 106–107.

Habits and occurrence of *Micropallas whitneyi, Dryobates arizonae, Eugenes fulgens,* and *Basilinna leucotis.*

French, N. R.

1959. Distribution and migration of the Black Rosy Finch. Condor 61: 18–29.

Arizona records are listed.

Friedmann, H.

1927. A revision of the classification of the Cowbirds. Auk 44: 495–508.

Ranges of *Molothrus ater obscurus* and *Tangavius aeneus aeneus* include Arizona.

———

1947. Geographic variations of the Black-bellied, Fulvous, and White-faced Tree-ducks. Condor 49: 189–195.

Gives characters and ranges of *Dendrocygna autumnalis lucida* and *D. bicolor helva,* both listed from Arizona.

Garth, J. S.

1943. [Report of Lark Buntings at Gila Bend.] [In] Minutes of Cooper Club meetings, Condor 45: 43.

George, W.

1958. Records of eastern birds from the Chiricahua Mountains of Arizona. Auk 75: 357–359.

An account of 8 species.

Gibbs, R. H., Jr. and S. P. Gibbs

1956. Rose-throated Becard nesting in the Chiricahua Mountains, Arizona. Wilson Bull. 68: 77–78.

Gibson, F.

1946. The Great-tailed Grackle in Arizona. Ariz. Wildlife and Sportsman, 7 (8): 10.

At San Carlos.

Gilman, M. F.

1914. Breeding of the Bronzed Cowbird in Arizona. Condor 16: 255–259.

At Sacaton and Santan.

———

1914. Notes from Sacaton, Arizona. Condor 16: 260–261.

Adds three species to the Arizona list: *Zonotrichia querula, Sphyrapicus varius daggetti,* and *Aythya collaris.*

Gilman, M. F.
1915. A forty acre bird census at Sacaton, Arizona. Condor 17: 86–90.
Notes on 21 species, most of them nesting.

Girard, G. L.
1937. Life history, habits, and food of the Sage Grouse, *Centrocercus urophasianus* Bonaparte. Univ. Wyo. Publ. 3 (1): 1–56.
Original distribution "probably" included, among other states, Arizona. No proof offered.

Godfrey, W. E.
1944. Five birds unusual in Arizona. Auk 61: 149–150. *Calypte anna, Colaptes auratus luteus, C. a. borealis, Sphyrapicus varius ruber,* and *Hylocichla minima minima.*

Goldman, E. A.
1902. In search of a new turkey in Arizona. Auk 19: 121–127.
Observations on *Meleagris gallopavo merriami* in the Mogollon Mountains.

———
1926. Breeding birds of a White Mountains lake. Condor 28: 159–164.
Accounts of 19 species from Marsh Lake.

Gollop, J. B.
1963. Autumnal distribution of young Mallards banded at Kindersley, Saskatchewan. In Proc. 13th Int. Ornith. Congr. 2: 855–865.
Lists one Arizona record.

Grater, R.
1934. Some migrating Sparrows and related species. Grand Canyon Nature Notes 9: 335–336.
At the Grand Canyon.

———
1934. Current observations on bird banding at Grand Canyon. Grand Canyon Nature Notes 9: 354.

———
1935. Some wildlife observations on the Canyon floor. Grand Canyon Nature Notes 9: 365–367.
Birds observed on a three-day trip in November into the Grand Canyon.

———
1935. Miscellaneous field notes. Grand Canyon Nature Notes 9: 382.
Accounts of three species of birds at the Grand Canyon.

———
1935. Miscellaneous field notes. Grand Canyon Nature Notes 9: 394.
Accounts of four species of birds at the Grand Canyon.

———
1935. Brewster's Egret at Grand Canyon National Park. Auk 52: 443.

———
1936. Bird banding from October 15, 1934, to October 15, 1935. Grand Canyon Nat. Hist. Assoc. Bull. 4: 1–4.
At Grand Canyon.

———

1936. Preliminary survey of the status of the Dusky Grouse on the Kaibab Plateau. Grand Canyon Nat. Hist. Assoc. Bull. 4: 7–8.

———

1936. Pileated Woodpecker in Grand Canyon National Park. Auk 53: 218.

———

1937. Check-list of birds of Grand Canyon National Park. Grand Canyon Nat. Hist. Assoc. Bull. 8: 1–55.

———

1938. Song Sparrow records from the Grand Canyon in northern Arizona. Wilson Bull. 50: 60–61.

———

1939. New bird records for Nevada. Condor 41: 30.
White-fronted Goose, Bonaparte Gull, and Least Tern on Lake Mead.

———

1939. New bird records for Clark County, Nevada. Condor 41: 220–221.
Accounts of species at Lake Mead and along the Colorado River.

———

1947. The birds of Zion, Bryce, and Cedar Breaks, Utah. Zion-Bryce Nat. Hist. Assoc. Mus. Bull. 5: 1–93.
Includes discussion of distribution of Red-backed Junco on the Kaibab Plateau.

Grinnell, J.
1914. An account of the mammals and birds of the lower Colorado Valley. Univ. Calif. Publ. Zool. 12 (4): 51–294.
Accounts of 150 species and subspecies of birds, with record of specimens, distribution, variation, and notes on behavior.

———

1917. An invasion of California by the Eastern Goshawk. Condor 19: 70–71.
One collected south of Palo Verde in the Colorado Valley.

———

1918. Seven new or noteworthy birds from east-central California. Condor 20: 86–90.
Includes a drawing of the bill of a specimen of *Sitta carolinensis nelsoni* from the Sierra Ancha, Arizona.

———

1923. The present state of our knowledge of the Gray Titmouse in California. Condor 25: 135–137.
Specimens examined from northern Arizona.

———

1926. A new race of the White-breasted Nuthatch from Lower California. Univ. Calif. Publ. Zool. 21 (15): 405–410.
Includes a drawing of the head of *S. c. nelsoni* from Gila County, Arizona.

Grinnell, J.

1927. Designation of a Pacific Coast subspecies of Chipping Sparrow. Condor 29: 81–82.

Measurements given of Arizona specimens of *S. p. arizonae.*

———

1927. A new race of Gila Woodpecker from Lower California. Condor 29: 168–169.

Compares Arizona specimens of *C. u. uropygialis.*

——— and H. S. Swarth

1913. An account of the birds and mammals of the San Jacinto area of southern California with remarks upon the behavior of geographic races on the margins of their habitats. Univ. Calif. Publ. Zool. 10 (10): 197–406.

Includes comments on *Dryobates scalaris cactophilus* and *Vireo vicinior* of Arizona.

Griscom, L.

1931. Notes on rare and little known neotropical Pygmy Owls. Proc. New England Zool. Club 12: 37–43.

Includes comments on *Glaucidium gnoma* from Arizona.

———

1937. A monographic study of the Red Crossbill. Proc. Boston Soc. Nat. Hist. 41 (5): 77–210.

An important study with numerous data on migrations and breeding seasons in Arizona.

Gross, A. O.

1921. The Dickcissel (*Spiza americana*) of the Illinois prairies. Auk 38: 163–184.

Lists a specimen from Tucson, Arizona, taken September 11, 1884.

———

1968. Albinistic eggs (white eggs) of some North American birds. Bird-Banding 39: 1–6.

Lists Red-tailed Hawk with one albino egg and one of normal coloration: Bent, A. C. (1922) Cochise, Arizona.

Gullion, G. W., W. M. Pulich, and F. G. Evenden

1959. Notes on the occurrence of birds in southern Nevada. Condor 61: 278–297.

Includes observations on Lake Mohave.

Guthrie, J. D.

1927. About wild Turkeys. Ariz. Wild Life 1 (2): 1–3.

In the White Mountains.

Haldeman, J. R.

1968. Breeding birds of a ponderosa pine forest and a fir, pine, aspen forest in the San Francisco Mountains, north-central Arizona. Abstract of paper read at 12th annual meeting of the Ariz. Acad. Sci. at Flagstaff. Jour. Ariz. Acad. Sci. 5, proc. suppl.: 17.

Halloran, A. F.
1947. Birds of Yuma County. Ariz. Wildlife and Sportsman 8 (11): 10–11, 33.

———
1948. Whistling Swans in the tules. Ariz. Wildlife and Sportsman 9 (11): 7.
On the Colorado River north of Yuma.

Hamilton, T. H.
1958. Adaptive variation in the genus *Vireo*. Wilson Bull. 70: 307–346.
Measurements of Arizona specimens are included.

Hansen, H. A.
1967. Waterfowl status report 1967. U.S. Dept. Int. Bur. Sport Fisheries and Wildlife. Spec. Sci. Rep. Wildl. 111. 144 pp.

——— and M. R. Hudgins
1966. Waterfowl status report 1966. U.S. Dept. Int. Bur. Sport Fisheries and Wildlife. Spec. Sci. Rep. Wildl. 99. 96 pp.

Hargrave, L. L.
1932. Banding report—1931. Museum of Northern Arizona bird banding station. News from the Bird-Banders 7 (2): 30.

———
1932. Woodhouse Jays on the Hopi Mesas, Arizona. Condor 34: 140–141.
Common at Hopi Indian villages.

———
1932. Miscellaneous bird notes from the San Francisco Mountains. Grand Canyon Nature Notes 7 (2): 18–21.
Records of 18 species.

———
1932. Notes on fifteen species of birds from the San Francisco Mountain region, Arizona. Condor 34: 217–220.
Loxia curvirostra bendirei is new to the Arizona list.

———
1932. The American Golden-eye in Arizona. Condor 34: 227.
Several records.

———
1933. Bird life of the San Francisco Mountains, Arizona. No. 1. Mus. N. Ariz. Mus. Notes 5 (10): 57–60.

———
1933. Some fall migration notes from northern Arizona lakes. Condor 35: 75–77.
Accounts of 13 species, of which 2 are new to the Arizona list: *Colymbus auritus* and *Larus californicus*.

———
1933. Bird life of the San Francisco Mountains, Arizona. No. 2: winter birds. Mus. N. Ariz. Mus. Notes 6 (6): 27–34.

Hargrave, L. L.
1933. Bird banding in northern Arizona. Part 1, Flagstaff station. Grand Canyon Nature Notes 7 (11): 114–116.
A total of 1,504 birds were banded.

———
1933. Some bird notes from Bly, Arizona. Grand Canyon Nature Notes 8 (6): 184–187.
Accounts of 7 species.

———
1935. An additional record of the Whistling Swan in Arizona. Condor 37: 177–178.
In the Flagstaff region.

———
1935. Nine new birds from Williams, Arizona. Condor 37: 285.

———
1936. Bird life of the San Francisco Mountains, Arizona. No. 4: Swans, Geese, and Ducks. Mus. N. Ariz. Mus. Notes 9 (5): 25–31.

———
1936. Stilt Sandpiper in Arizona. Auk 53: 211.
Three collected April 25, 1933, near Tempe.

———
1936. Seven birds new to Arizona. Condor 38: 120–121.
Limosa fedoa, Stercorarius pomarinus, Muscivora forficata, Otocoris alpestris lamprochroma, Hylocichla ustulata swainsoni, Dendroica aestiva rubiginosa, Melospiza melodia fisherella.

———
1936. New bird records for Arizona. Condor 38: 171–172.
Accounts of 10 species.

———
1939. Winter bird notes from Roosevelt Lake, Arizona. Condor 41: 121–123.
Records of numerous species.

———
1964. Additional records of the Scissor-tailed Flycatcher in Arizona. Condor 66: 438.
Two at Pomerene.

——— and A. R. Phillips
1936. Bird life of the San Francisco Mountains, Arizona. No. 3. Mus. N. Ariz. Mus. Notes 8 (9): 47–50.

———, A. R. Phillips, and R. Jenks
1937. The White-winged Junco in Arizona. Condor 39: 258–259.
First Arizona specimen from near Springerville; others from Flagstaff vicinity.

Harper, F.
1930. A historical sketch of Botteri's Sparrow. Auk 47: 177–185.
Brief history of the species in Arizona.

Harris, H.
1941. The annals of *Gymnogyps* to 1900. Condor 43: 3–55.
Sight record of the Condor at Fort Yuma in 1865.

Harrison, B. [= W. I.]
1960. [Winter bird-population study, at] Ranch pond in arid country [near Nogales, Ariz.] Aud. Field Notes 14: 356.

———
1961. [Winter bird-population study at] Ranch pond in arid country [near Nogales, Ariz.] Aud. Field Notes 15: 372.

Harrison, W. I.
1962. The first record of the Rufous-backed Robin in the United States. Auk 79: 271.
Near Nogales, Arizona.

Hasbrouck, E. M.
1893. The geographical distribution of the genus *Megascops* in North America. Auk 10: 250–264.
Includes *Megascops asio trichopsis* and *M. flammeolus* in Arizona.

Hawkins, A. S.
1949. Migration pattern in the Mallard, pp. 4–5, in J. W. Aldrich and others, Migration of some North American waterfowl. U.S. Dept. Int. Fish and Wildl. Serv. Spec. Sci. Rep. (Wildl.) no. 1.
Map of recoveries includes Arizona.

Hayward, C. L.
1967. Birds of the Upper Colorado River Basin. Brigham Young Univ. Sci. Bull. Biol. Ser. 9 (2): 1–64.
Contains a few northern Arizona records.

Heald, W. F.
1952. Nature's crossroads: the Chiricahua Mountains. Aud. Mag. 54: 298–304, 340.
Includes comments on some of the birds.

Heaton, L.
1939. Movements of Gambel Sparrows at Pipe Spring, Arizona. News from the Bird-Banders 14 (4): 44.

Heermann, A. L.
1859. Report upon the birds collected on the survey. In Zoological Report, Explorations and Surveys for a railroad route from the Mississippi River to the Pacific Ocean. Vol. 10, no. 1: 9–20, plus one unnumbered page.
List of 25 species, not all from Arizona.

———
1859. Report upon birds collected on the survey. In Report of Explorations in California for railroad routes. In Explorations and Surveys for a railroad route from the Mississippi River to the Pacific Ocean. Vol. 10, part 4, no. 2: 29–80.
Contains some notes from Fort Yuma.

Hein, D.
1969. Commentary on the 1968 annual report. Western Bird Bander 44: 23.
Includes banding summary, 1960 to 1968; Arizona and New Mexico data are combined.

Henderson, J.
1908. The Mountain Bluebird in northern Arizona. Condor 10: 94.
Suspects that small numbers of this species may remain in winter in the mountains.

Henshaw, H. W.
1874. On a Hummingbird new to our fauna, with certain other facts ornithological. Amer. Nat. 8: 241–243.
Eugenes fulgens near Camp Grant.

———
1875. Annotated list of the birds of Arizona. Pp. 153–166. In App. I 2 of App. LL of Annual Report Geog. Exp. and Surv. W. 100th Merid., etc. (Wheeler).
A list of 294 species (erroneously numbered 291).

———
1876. On two *Empidonaces, traillii* and *acadicus.* Bull. Nuttall Ornith. Club 1: 14–17.
Mentions range of *E. t. pusillus* in Arizona.

———
1877. Description of a new species of Hummingbird from California. Bull. Nuttall Ornith. Club 2: 53–58.
An Arizona specimen of *Selasphorus rufus* is listed.

———
1884. The Shore Larks of the United States and adjacent territory. Auk 1: 254–268.
Includes a brief discussion of specimens from Arizona.

Hespenheide, H. A.
1964. Competition and the genus *Tyrannus.* Wilson Bull. 76: 265–281.
Maps of ranges of four species in Arizona; observations in southern Arizona.

[Hickey, M. B., ed.]
1947. The distribution of North American birds. Aud. Field Notes 1: 132–134.
The map of breeding localities of the Lark Sparrow (*Chondestes grammacus*) in the United States includes Arizona.

Hoffmeister, D. F.
1966. Records of northern Arizona mammals. Plateau 39: 90–93.
Mentions a Long-eared Owl shot near San Carlos, Gila County, Arizona.

Hollister, N.
1908. Birds of the region about Needles, California. Auk 25: 455–462.
Notes on several species observed on the Arizona side of the Colorado River at Fort Mohave.

Holterhoff, E., Jr.
1881. A collector's notes on the breeding of a few western birds. Amer. Nat. 15: 208–219.
Includes accounts of several species in the Tucson area.

Howard, O. W.
1899. [Great Blue Herons on the San Pedro River.] Bull. Cooper Ornith. Club 1: 55.
Albino heron observed.

———
1906. The English Sparrow in the southwest. Condor 8: 67–68.
Observed at Tucson and Tombstone.

Howell, A. B.
1916. Some results of a winter's observations in Arizona. Condor 18: 209–214.
A running account of birds observed at Fort Lowell from Dec. 7, 1915 to March 25, 1916.

——— and A. van Rossem
1915. Additional observations on the birds of the lower Colorado Valley in California. Condor 17: 232–234.
Remarks on Arizona species.

Howell, T. R.
1951. Lapland Longspur in Arizona. Condor 53: 262.
In the Petrified Forest National Monument.

———
1952. Natural history and differentiation in the Yellow-bellied Sapsucker. Condor 54: 237–282.
Maps of breeding and winter ranges in Arizona; discussion of variation and intergradation.

———
1953. Racial and sexual differences in migration in *Sphyrapicus varius.* Auk 70: 118–126.
Maps of breeding and winter ranges include Arizona.

Hubbard, J. P.
1965. The summer birds of the forests of the Mogollon Mountains, New Mexico. Condor 67: 404–415.
Incidental mention of a Crossbill collected in the Chiricahua Mountains, Arizona.

———
1969. The relationships and evolution of the *Dendroica coronata* complex. Auk 86: 393–432.
Includes Arizona specimens.

Huey, L. M.
1920. Two birds new to the Lower Colorado River region. Condor 22: 73.
Lophodytes cucullatus and *Toxostoma curvirostre palmeri* collected near Bard on the California side of the river.

———
1926. Two species new to the avifauna of California. Condor 28: 44.
Junco mearnsi and *Dendroica tigrina* collected along the lower Colorado River.

———
1926. Notes from northwestern Lower California, with the description of an apparently new race of the Screech Owl. Auk 43: 347–362.
Specimens examined include *O. a. cineraceus* from Arizona localities.

Huey, L. M.
1930. Notes from the vicinity of San Francisco Mountain, Arizona. Condor 32: 128.
Accounts of Arizona Spotted Owl, Red-faced Warbler, and Painted Redstart.

———
1931. The most western record of the Indigo Bunting. Condor 33: 129.
In the Huachuca Mountains.

———
1933. White Mountain Fox Sparrow in Arizona. Condor 35: 204.
Collected at Big Sandy Creek.

———
1935. The Charleston Mountain Blue-fronted Jay at Castle Dome, Yuma County, Arizona. Condor 37: 257.

———
1936. Notes on the summer and fall birds of the White Mountains, Arizona. Wilson Bull. 48: 119–130.
A list of 69 species.

———
1936. Notes from Maricopa County, Arizona. Condor 38: 172.
Accounts of a number of winter birds in the vicinity of Gila Bend.

———
1939. Birds of the Mount Trumbull region, Arizona. Auk 56: 320–325.
Accounts of 51 species observed from July 24 to August 5, 1937.

———
1942. A vertebrate faunal survey of the Organ Pipe Cactus National Monument, Arizona. Trans. San Diego Soc. Nat. Hist. 9: 353–376.
Includes a list of 150 species and subspecies of birds.

———
1959. The second occurrence of a Brown Booby near Parker Dam on the Colorado River. Condor 61: 223–224.

———
1963. October notes from Willow Beach, Colorado River, Arizona. Jour. Ariz. Acad. Sci. 2: 108–112.
Accounts of 42 species.

Hunt, D. B.
1967. Species leaders for 1966. Western Bird Bander 42: 34–37.
A list of the largest number of species of birds banded in the western states.

Imler, R. H. and E. R. Kalmbach
1955. The Bald Eagle and its economic status. U.S. Dept. Int. Fish and Wildlife Serv. Circ. 30: 1–51.
No change in abundance on Colorado River refuges since the 1940s.

Jackson, H. H. T.
1922. Some birds of Roosevelt Lake, Arizona. Condor 24: 22–25.
Accounts of 13 species.

Jacot, E. C.
1931. Notes on the Spotted and Flammulated Screech Owls in Arizona. Condor 33: 8–11.
In the Huachuca Mountains.

———
1932. *Junco hyemalis connectens* in Arizona. Condor 34: 140.
Near Prescott.

———
1934. An Arizona nest of the Ferruginous Rough-leg. Condor 36: 84–85.
Near Prescott.

Jenkins, H. O.
1906. Variation in the Hairy Woodpecker (*Dryobates villosus* and subspecies). Auk 23: 161–171.
Map of distribution.

Jenks, R.
1931. Birding in Grand Canyon. Grand Canyon Nature Notes 5: 75–77.
Birds observed on a trip to Phantom Ranch.

———
1931. Some birds and their nests on the Kaibab. Grand Canyon Nature Notes 5: 101–105.
Nests of 16 species recorded; 48 species observed.

———
1932. Ornithology of the life zones. Summit of San Francisco Mts. to bottom of Grand Canyon. U.S. Dept. Int. Nat. Park Serv. Tech. Bull. 5, 31 pp., with page of errata not numbered.
A list of the birds.

———
1934. Unusual nesting records from northern Arizona. Condor 36: 172–176.
Accounts of 8 species.

———
1936. Two new records for Arizona. Condor 38: 38.
Spinus psaltria psaltria and *Molothrus ater artemisiae* from vicinity of Springerville.

——— and J. O. Stevenson
1935. Breeding records of the Catbird in Arizona. Condor 37: 81.
Near Springerville.

——— and J. O. Stevenson
1937. Northern Arizona bird notes. Condor 39: 40–41.
Accounts of 6 species from the San Francisco Mountain region.

——— and J. O. Stevenson
1937. Bird records from central-eastern Arizona. Condor 39: 87–90.
Accounts of 24 species.

——— and J. O. Stevenson
1937. Notes on range of Bendire Thrasher in Arizona and New Mexico. Condor 39: 126.
Additional records from northeastern Arizona.

Jensen, G. H.
1949. Migration of the Gadwall, pp. 9–10, with map, in Migration of some North American Waterfowl. U.S. Dept. Int. Fish and Wildlife Serv. Spec. Sci. Rep. (Wildl.) no. 1.

———
1949. Migration of the Canada Goose, p. 47, with map, in Migration of some North American Waterfowl. U.S. Dept. Int. Fish and Wildlife Serv. Spec. Sci. Rep. (Wildl.) no. 1.

Jeter, H. H.
1959. Cliff Swallows of mixed plumage types in a colony in southeastern Arizona. Condor 61: 434.
Along the San Pedro River.

Johnson, D. H., M. D. Bryant, and A. H. Miller
1948. Vertebrate animals of the Providence Mountains area of California. Univ. Calif. Publ. Zool. 48 (5): 221–376.
Comparisons made with several species of Arizona birds.

Johnson, N. K.
1963. The supposed migratory status of the Flammulated Owl. Wilson Bull. 75: 174–178.
Believed to be nonmigratory: possibility of torpidity is suggested.

———
1965. The breeding avifaunas of the Sheep and Spring Ranges in southern Nevada. Condor 67: 93–124.
Contains brief mention of some Arizona species.

———
1965. Differential timing and routes of the spring migration in the Hammond Flycatcher. Condor 67: 423–437.
Includes many Arizona specimens.

Johnson, R. R. and B. Roer
1968. Changing status of the Bronzed Cowbird in Arizona. Condor 70: 183.
Reports increase in numbers and range extension northward.

Jollie, M.
1947. Plumage changes in the Golden Eagle. Auk 64: 549–576.
Specimens from Arizona mentioned.

Jones, J. C.
1940. Food habits of the American Coot, with notes on distribution. U.S. Dept. Int. Bur. of Biol. Serv. Wildl. Research Bull. 2: 1–52.
Contains map of breeding and winter ranges.

Jones, L. and W. L. Dawson
1900. A summer reconnaissance in the west. Wilson Bull. 33: 1–39.
Birds are listed from the Flagstaff area and from Mellon on the Colorado River.

Judson, W. B.
1897. [Nests of the White-throated Swift in the Huachuca Mountains.] Nidiologist 4: 91–92.

Kalmbach, E. R.
1927. The Magpie in relation to agriculture. U.S. Dept. Agric. Tech. Bull. 24: 1–29.
The distributional map includes parts of northern Arizona.

Kartchner, K. C.
1932. Wild fowl breeding in northern and eastern Arizona. Ariz. Wild Life 4 (3): 8.
Comments on scarcity of ducks.

Kassel, H. L.
1941. Winter birds on the campus at Flagstaff. Plateau 13: 65–68.

Keith, A. R.
1968. A summary of the extralimital records of the Varied Thrush, 1848 to 1966. Bird-Banding 39: 245–276.
Lists three Arizona records.

Kellogg, R. T.
1922. Rare birds in Arizona and New Mexico. Condor 24: 29–30.
Parabuteo unicinctus harrisi collected near Mesa, Arizona.

Kelso, L.
1934. A key to species of American Owls. Biol. Leaflet no. 4: 1–101. Intelligencer Printing Co.
Includes brief accounts of distribution of Arizona species.

Kennard, F. H.
1924. Some Arizona notes. Condor 26: 76–77.
Accounts of 8 species in southern Arizona; *Passerella iliaca canescens* collected at Oracle, new to the Arizona list.

Kennerly, C. B. R.
1856. Field notes and explanations. In Report on the Zoology of the Expedition. In Explorations and Surveys for a railroad route from the Mississippi River to the Pacific Ocean. Vol. 4, part 6, no. 1: 1–17.
List of specimens of birds collected.

———
1859. Report on birds collected on the route. In Zoological Report. In Explorations and Surveys for a railroad route from the Mississippi River to the Pacific Ocean. Vol. 10, part 6, no. 3: 19–35.
Lists 88 species of birds, including some from Texas, New Mexico, and California.

Kerr, R. M.
1966. Quack comeback. Our Public Lands 16 (2): 4–5.
The population of the New Mexican Duck is estimated to be about 20 in Arizona.

Kessel, B.
1953. Distribution and migration of the European Starling in North America. Condor 55: 49–67.
Includes map of Arizona records.

Kiel, W. H., Jr.
1960. Mourning Dove status report, 1960. U.S. Dept. Int. Bur. Sport Fisheries and Wildlife, Spec. Sci. Rep.—Wildl. 49: 1–34.

Kiel, W. H., Jr.
1961. Mourning Dove status report, 1961. U.S. Dept. Int. Bur. Sport Fisheries and Wildlife, Spec. Sci. Rep.—Wildl. 57: 1–34.

Kimball, H. H.
1921. Notes from southern Arizona. Condor 23: 57–58.
Accounts of birds observed at Tucson, the Chiricahua Mountains, Willcox, and Yuma.

———
1922. Bird records from California, Arizona, and Guadalupe Island. Condor 24: 96–97.
Accounts of five species from Arizona.

———
1923. Bird notes from Arizona and California. Condor 25: 109.
Wood Ibis, Black Vulture, and Western Willet observed near Tucson.

King, J. R., D. S. Farner, and L. R. Mewaldt
1965. Seasonal sex and age ratios in populations of the White-crowned Sparrows of the race *gambelii*. Condor 67: 489–504.
Includes Arizona winter populations.

Kirk, R.
1955. A year among the cactus giants. Aud. Mag. 57: 18–21.
Incidental mention of a few birds in the Organ Pipe Cactus National Monument.

[Kirsher, W. K.]
1965. Some recoveries included in annual reports. Western Bird Bander 40: 32.
Mourning Dove banded in Nevada recovered in Tucson.

Kravits, I.
1934. [Yellow-headed Blackbird at Grand Canyon.] Grand Canyon Nature Notes 9: 282.

Lanyon, W. E.
1960. The Middle American populations of the Crested Flycatcher *Myiarchus tyrannulus*. Condor 62: 341–350.
Arizona specimens are discussed.

Law, J. E.
1917. Notes on the Arizona Spotted Owl. Condor 19: 69.
Specimens from the Chiricahua Mountains.

[Law, J. E.]
1918. [Report of Thick-billed Parrots in Arizona.] Condor 20: 100.

Law, J. E.
1924. California Pelican: An addition to the Arizona List. Condor 26: 153.
At Dos Cabezas, Cochise County, Arizona.

———
1928. *Toxostoma curvirostris:* I. Description of a new subspecies from the lower Rio Grande. Condor 30: 151–152.
Specimens from the Chiricahua Mountains are *T. c. curvirostris*.

Lee, M. H.
1920. Notes on a few birds of the Grand Canyon, Arizona. Condor 22: 171–172.
Accounts of 8 species.

Lensink, C. J.
1964. Distribution of recoveries from bandings of ducklings. U.S. Dept. Int. Fish and Wildl. Serv. Bur. Sport Fisheries and Wildlife. Spec. Sci. Rep. Wildlife no. 89: 1–146.
Includes Arizona recoveries.

Leopold, A.
1942. A raptor tally in the northwest. Condor 44: 37–38.
Includes northern Arizona.

Levy, S. H.
1958. A new bird record for Arizona. Condor 60: 70.
A specimen of Semipalmated Sandpiper from near Sasabe.

———
1958. A new United States nesting area for the Rose-throated Becard. Auk 75: 95.
In Guadalupe Canyon, Cochise County, Arizona.

———
1958. A possible United States breeding area for the Violet-crowned Hummingbird. Auk 75: 350.
In Guadalupe Canyon, Cochise County, Arizona.

———
1959. Thick-billed Kingbird in the United States. Auk 76: 92.
In Guadalupe Canyon, Cochise County, Arizona.

———
1959. The Least Grebe recorded again in Arizona. Condor 61: 226.
One collected near Sasabe.

———
1961. Two new birds recorded for Arizona. Condor 63: 98.
Ardea herodias herodias from Picacho Reservoir, and *Buteo harlani* from vicinity of St. David.

———
1961. The Caracara nesting in Arizona. Auk 78: 99.
On the Papago Indian Reservation.

———
1962. The first record of the Fan-tailed Warbler in the United States. Auk 79: 119–120.
In the Guadalupe Mountains, Cochise County, Arizona.

———
1962. The Ridgway Whip-poor-will in Arizona. Condor 64: 161–162.
In Guadalupe Canyon.

———
1964. What has happened to the Mexican Duck? Aud. Field Notes 18: 558–559.
Its habitat in the San Simon Cienega is threatened.

Ligon, J. D. and R. P. Balda
1968. Recent data on summer birds of the Chiricahua Mountains area, southeastern Arizona. Trans. San Diego Soc. Nat. Hist. 15 (5): 41–50.
They list 167 species; *Sialia sialis* is now a regular breeder in the area.

Ligon, J. S.
1952. The vanishing Masked Bobwhite. Condor 54: 48–50.
Brief notes on the attempt to introduce the species in Arizona.

———
1961. New Mexico birds, and where to find them. Univ. New Mexico Press. 360 pp.
Contains a number of references to bird distribution in Arizona.

Lincoln, F. C.
1917. Some notes of the birds of Rock Canyon, Arizona. Wilson Bull. 29: 65–73.
Contains remarks on 66 species and subspecies of birds from the vicinity of the Santa Catalina Mountains.

———
1923. The White Ibis in California. Condor 25: 181.
Lists a sight record by J. Hornung near the Colorado River.

———
1927. Returns from banded birds 1923 to 1926. U.S. Dept. Agric. Tech. Bull. 32: 1–96.
Includes Arizona species.

———
1927. Status of the Yellow-legs in Arizona. Condor 29: 164–165.
Totanus flavipes collected by E. A. Mearns in 1892 at the San Bernardino Ranch in southeastern Arizona.

———
1942. Waterfowl banding. Ariz. Wildlife and Sportsman 4 (6): 5.
A letter reporting that a Canada Goose recovered [in Arizona] was banded in Utah.

———
1950. Migration of birds. U.S. Dept. Int. Fish and Wildlife Serv. Circ. 16: 1–102.
Contains maps of migration routes of several species of Arizona birds.

Linsdale, J. M.
1936. The birds of Nevada. Pac. Coast Avi. 23: 1–145.
Contains incidental mention of birds along the Colorado River.

———
1937. The natural history of Magpies. Pac. Coast Avi. 25: 1–234.
Contains a summary of Arizona range of *Pica pica hudsonia*.

———
1957. Goldfinches on the Hastings Natural History Reservation. Amer. Mid. Nat. 57: 1–119.
Includes summary of Arizona distribution of *Spinus lawrencei*.

Loomis, L. M.
1901. An addition to the A. O.U. Check-List. Auk 18: 109–110.
Dendroica nigrifrons collected in the Huachuca and Chiricahua mountains.

Low, S. H.
1949. The migration of the Pintail, pp. 13–16, with 4 maps, in Migration of some North American Waterfowl. U.S. Dept. Int. Fish and Wildl. Serv. Spec. Sci. Rep. (Wildl.) no. 1.
They cross Arizona on their way elsewhere.

———
1949. Migration of the Green-winged Teal, pp. 17–18, with 3 maps, in Migration of some North American Waterfowl. U.S. Dept. Int. Fish and Wildlife Serv. Spec. Sci. Rep. (Wildl.) no. 1.

Lowery, G. H., Jr. and R. J. Newman
1966. A continentwide view of bird migration on four nights in October. Auk 83: 547–586.
Two Arizona stations are listed, but no migration data are included.

Lusk, R. D.
1899. New nesting location of Rivoli Hummer *(Eugenes fulgens).* Osprey 3: 140–141.
In the Huachuca Mountains, Arizona.

———
1900. Parrots in the United States. Condor 2: 129.
Rhynchopsitta pachyrhyncha observed in the Chiricahua Mountains.

———
1901. In the summer home of the Buff-breasted Flycatcher. Condor 3: 38–41.
Observations in the Santa Rita and Chiricahua mountains.

———
1921. The White-eared Hummingbird in the Catalina Mountains, Arizona. Condor 23: 99.

Marshall, J. T., Jr.
1956. Summer birds of the Rincon Mountains, Saguaro National Monument, Arizona. Condor 58: 81–97.
Account of 71 species with an extensive discussion of their environmental preferences.

———
1957. Birds of pine-oak woodland in southern Arizona and adjacent Mexico. Pac. Coast Avi. 32: 1–125.
An extensive account of the vegetation and of habitat preferences of the birds.

Martin, P. S.
1961. Southwestern animal communities in the late Pleistocene. Pp. 56–66. in L. M. Shields and J. L. Gardner, eds., Bio-ecology of the arid and semi-arid lands of the Southwest. Symposium held at New Mexico Highlands Univ. Las Vegas, New Mexico, 1958. New Mexico Highlands Univ. Las Vegas, New Mexico. 69 pp.
Gives altitudinal records of 85 breeding birds of the Chiricahua Mountains, Arizona.

Martinson, R. K., J. F. Voelzer, and M. R. Hudgins
1968. [1969]. Waterfowl status report, 1968. U.S. Dept. Int. Bur. Sport Fisheries and Wildl. Spec. Sci. Rep.—Wildl. 122: 1–158.

———, J. F. Voelzer, and S. L. Meller
1969. Waterfowl status report, 1969. U.S. Dept. Int. Bur. Sport Fisheries and Wildl. Spec. Sci. Rep.—Wildl. 128: 1–153.

McCaskie, G.
1965. The Curve-billed Thrasher in California. Condor 67: 443–444.
Includes records along the west bank of the Colorado River.

———
1966. The occurrence of Longspurs and Snow Buntings in California. Condor 68: 597–598.
Calcarius ornatus reported from the lower Colorado River.

——— and E. A. Cardiff
1965. Notes on the distribution of the Parasitic Jaeger and some members of the Laridae in California. Condor 67: 542–544.
Larus pipixcan observed at Imperial Dam on the Colorado River.

———, R. Stallcup, and P. DeBenedictis
1966. Notes on the distribution of certain Icterids and Tanagers in California. Condor 68: 595–597.
Cassidix mexicanus and *Tangavius aeneus* reported from the lower Colorado River.

McClanahan, R. C.
1940. Original and present breeding ranges of certain game birds in the United States. U.S. Dept. Int. Bur. Biol. Surv. Wildl. Leaflet BS-158: 1–21.
Contains numerous maps.

McClure, H. E.
1949. [Reports that the Black Phoebe banded on Kern River, California, was shot in Arizona.] News from the Bird-Banders 24: 28.

McGee, W. J.
1910. Notes on the Passenger Pigeon. Science, N. S. 32 (835): 958–964.
Nesting abundantly at Tinajas Altas! A misidentification.

McGregor, R. C.
1900. A list of unrecorded albinos. Condor 2: 86–88.
Agelaius phoeniceus from Phoenix, Arizona.

McHenry, D. E.
1932. [Shrikes at Bright Angel Point.] Grand Canyon Nature Notes 7: 76.

McKee, B. H.
1931. A lively little beggar. Grand Canyon Nature Notes 5: 56–57.
Notes on the Gray Titmouse at Grand Canyon.

McKee, E. D.
1927. Birds of the Havasupai Canyon. Grand Canyon Nature Notes 2: 2.
A list of 20 species.

McKee, E. D.
1927. Trailside nests. Grand Canyon Nature Notes 2: 3–4.
Nests of Black-throated Gray Warbler, Warbling Vireo, and White-throated Swift.

———
1928. Bird life on the Tonto Platform. Grand Canyon Nature Notes 3 (1): 2–3.

———
1928. New bird records for park. Grand Canyon Nature Notes 3 (3): 2.
From Kaibab Forest: Brown Creeper, Western Lark Sparrow, Marsh Hawk.

———
1929. The Horned Owl. Grand Canyon Nature Notes 3 (11): 3.

[McKee, E. D.]
1929. Feathered friends. Grand Canyon Nature Notes 4 (3): 15–16.
Brief accounts of 21 species.

———
1930. The Canyon Towhee. Grand Canyon Nature Notes 4 (11): 80.
Observed on the South Rim.

McKee, E. D.
1930. Geological and wildlife observations between Bass Canyon and Hermit Canyon. Grand Canyon Nature Notes 5 (1): 5–9.
A list of nine species of birds.

———
1930. Small but mighty! Grand Canyon Nature Notes 5 (2): 12.
Birds observed at a feeding station on the south rim of the Grand Canyon.

———
1931. [American Coot and Plumbeous Vireo at Grand Canyon.] Grand Canyon Nature Notes 5 (6): 61.

———
1931. [Aiken Screech Owl and White-throated Swift.] Grand Canyon Nature Notes 5 (7): 69–70.

———
1931. [Bird records for May at Grand Canyon.] Grand Canyon Nature Notes 5 (8): 83.
A list of 14 species.

———
1931. A quarrel among bird families. Grand Canyon Nature Notes 5 (11): 117.
Red-backed Juncos and Audubon Warblers.

———
1931. Additional notes on the Tanner Trail trip. Grand Canyon Nature Notes 5 (12): 121–122.
A list of 18 species of birds recorded.

———
1931. [Ruby-crowned Kinglets at Bright Angel Creek.] Grand Canyon Nature Notes 6 (2): 22.

McKee, E. D.
1932. [Arizona Spotted Owl at Grand Canyon.] Grand Canyon Nature Notes 6 (3): 28.

———
1932. [Bird notes from the Grand Canyon.] Grand Canyon Nature Notes 6 (5): 47.

———
1932. Some notes on bird feeding and photography. Grand Canyon Nature Notes 7 (1): 3–5.
At the Grand Canyon.

———
1932. [Scott Sparrow and Rocky Mountain Creeper.] Grand Canyon Nature Notes 7 (1): 10.

———
1932. Bird migration dates. Grand Canyon Nature Notes 7 (3): 28.
Arrival dates of 11 species are given.

———
1932. Recent bird records from Havasu Canyon. Grand Canyon Nature Notes 7 (5): 53.
Birds observed by R. Jenks.

———
1932. [Pygmy Nuthatches and Audubon's Hermit Thrush at the Grand Canyon.] Grand Canyon Nature Notes 7 (5): 54.

———
1932. Recent bird notes. Grand Canyon Nature Notes 7 (9): 91–92.
Accounts of 12 species from the Grand Canyon National Park.

———
1933. The hermit of Horn Creek. Grand Canyon Nature Notes 7 (11): 116–118.
A summary of Spotted Owl occurrences in northern Arizona.

———
1933. Bird banding in northern Arizona. Part II, Grand Canyon station. Grand Canyon Nature Notes 7 (12): 121–123.
Birds banded total 540.

———
1933. Junco visits both rims. Grand Canyon Nature Notes 8 (3): 157.

———
1933. [Mearn's Woodpeckers, Band-tailed Pigeons, and Marsh Hawks.] Grand Canyon Nature Notes 8 (7): 205.

———
1934. [Belted Kingfisher on the South Rim of the Grand Canyon.] Grand Canyon Nature Notes 9 (3): 290.

———
1936. Bird observations in Grand Canyon Nat'l Park from October 1, 1934 to October 1, 1935. Grand Canyon Nat. Hist. Assoc. 4: 9–17.

———

1938. [Report of returns of Rocky Mountain Nuthatch and Red-backed Junco.] News from the Bird-Banders 13: 46.
At Grand Canyon.

———

1939. Four species new to Grand Canyon National Park. Condor 41: 256–257.
Spizella arborea ochracea, Glaucidium gnoma pinicola, Phainopepla nitens lepida, and *Cryptoglaux acadica acadica.*

———

1942. Results from a bird banding station at Grand Canyon. Plateau 15: 10–13.

———

1945. Oak Creek Canyon. Plateau 18: 25–32.
Contains records of Water Ouzel and Band-tailed Pigeon.

McLean, D. D.
1969. Some additional records of birds in California. Condor 71: 433–434.
Myiarchus tyrannulus magister nesting north of Needles, San Bernardino County, California, near the Colorado River.

McMurry, F. M. [=B.]
1945. Bird observations on Lake Laguna. Ariz. Wildlife and Sportsman 6 (13): 10–13.
Notes on water birds on Mittry Lake.

McMurry, F. B.
1948. Brewster's Booby collected in the United States. Auk 65: 309–310.
On the Colorado River, at Imperial Dam, on the California side.

——— and G. Monson
1947. Least Grebe breeding in California. Condor 49: 125–126.
At the west end of Imperial Dam on the Colorado River.

Mearns, E. A.
1886. Some birds of Arizona. Zone-tailed Hawk. *Buteo abbreviatus* Caban. Mexican Black Hawk. *Urubitinga anthracina* (Licht.) (Lafr.) Auk 3: 60–73.
Includes descriptions and habits of both hawks.

———

1886. Some birds of Arizona. Genus *Harporhynchus* Cabanis. Auk 3: 289–307.
Synonymy, descriptions, and habits of *H. crissalis* and *H. lecontei.*

———

1890. Observations on the avifauna of portions of Arizona. Auk 7: 45–55.
Annotated list of species from the high mountains of central Arizona.

———

1890. Observations on the avifauna of portions of Arizona. Auk 7: 251–264.
Annotated list of species of birds from the high mountains of central Arizona.

Mearns, E. A.
1901. An addition to the avifauna of the United States. Proc. Biol. Soc. Wash. 14: 177–178.
Petrochelidon melanogaster collected along the San Bernardino and Santa Cruz rivers.

———
1902. Two subspecies which should be added to the Check-List of North American birds. Auk 19: 70–72.
Mimus polyglottos leucopterus from Arizona.

———
1902. Description of a hybrid between the Barn and Cliff Swallows. Auk 19: 73–74.
P. melanogaster occurs along the Mexican boundary from the San Luis Mountains to Nogales.

———
1911. Description of a new subspecies of the Painted Bunting from the interior of Texas. Proc. Biol. Soc. Wash. 24: 217–218.
Passerina ciris pallidior ". . . migrating to Arizona. . . ."

Mengel, R. M.
1964. The probable history of species formation in some northern Wood Warblers (Parulidae). The Living Bird, 3rd annual: 9–43.
Includes maps showing distribution in Arizona.

———
1890. List of birds noted at the Grand Canon of the Colorado, Arizona, September 10 to 15, 1889. N. Amer. Fauna 3: 38–41.

———
1890. Annotated list of birds of the San Francisco Mountain Plateau and the desert of the Little Colorado River, Arizona. N. Amer. Fauna 3: 87–101.

Merriam, C. H.
1895. The Leconte Thrasher, *Harporhynchus lecontei.* Auk 12: 54–60.
Distribution and nesting.

Mershon, W. B.
1919. The Whistling Swan in Arizona. Condor 21: 126.
Near Williams.

Messinger, N. G.
1967. Two June records of the Canada Goose in Grand Canyon, Arizona. Condor 69: 319.

———
1968. A Hairy Woodpecker from Petrified Forest National Park, Arizona. [In] A. H. Schroeder, ed. Collected papers in honor of Lyndon Lane Hargrave. Papers Archeol. Soc. New Mex. 1: 163–164.
Four observed.

Miller, A. H.
1928. The status of the Cardinal in California. Condor 30: 243–245.
R. c. superba occurs in Arizona.

———

1932. The summer distribution of certain birds in central and northern Arizona. Condor 34: 96–99.
Accounts of 34 species.

———

1936. The identification of Juncos banded in the Rocky Mountains States. Bird-Lore 38: 429–433.
Brief distributional notes on Arizona species.

———

1937. Biotic associations and life-zones in relation to the Pleistocene birds of California. Condor 39: 248–252.
Contains remarks on the distribution of the Bendire Thrasher in Arizona.

———

1947. Arizona race of Acorn Woodpecker vagrant in California. Condor 49: 171.
Mentions its occurrence in the Hualpai Range.

Miller, L.
1927. The Painted Redstart as a California bird. Condor 29: 77.
Also observed in the Huachuca Mountains, Arizona.

———

1929. The Elf Owl in western Arizona. Condor 31: 252–253.
In the Tumacacori Mountains and near Tucson.

———

1930. Further notes on the Harris Hawk. Condor 32: 210–211.
Nesting in the vicinity of Potholes, on the California side of the Colorado River.

———

1932. The Summer Tanager again in California. Condor 34: 48–49.
Mentions specimens collected in Arizona.

[Miller, L.]
1936. [Swainson Hawks observed in Arizona.] Condor 38: 93.

Miller, L.
1936. The Flammulated Screech Owl on Mount Pinos. Condor 38: 228–229.
Includes remarks on its occurrence in the Chiricahua Mountains, Arizona.

[Miller, L.]
1940. [Report of a peculiar Sparrow Hawk in Arizona.] Condor 42: 268.

Miller, L.
1946. The Elf Owl moves west. Condor 48: 284–285.
Notes on Arizona distribution and habits.

———

1957. Some avian flyways of western America. Wilson Bull. 69: 164–169.
Mentions the Santa Cruz Valley and Sycamore Canyon, Arizona, as flyways.

———, W. P. Taylor, and H. S. Swarth
1929. Some winter birds at Tucson, Arizona. Condor 31: 76–77.
The list of numerous species includes Black Vulture and Gray Titmouse.

Miller, W. De W.
1910. The Red-billed Tropic-bird in Arizona. Auk 27: 450–451.
Breninger's Yellow-billed Tropic-bird (Auk 22: 408, 1905) proves to be *Phaëthon aethereus*.

Moisan, G.
1967. The Green-winged Teal: its distribution, migration, and population dynamics. Bur. Sport Fisheries and Wildlife. Spec. Sci. Rep.—Wildl. 100: 1–248.
Includes banding recoveries in Arizona.

Monson, G.
1936. The Great-tailed Grackle in Arizona. Wilson Bull. 48: 48.
Near Safford.

———
1936. Bird notes from the Papago Indian Reservation, southern Arizona. Condor 38: 175–176.
Accounts of 36 species observed from September 1, 1934, to February 15, 1935.

———
1936. Nesting of the Black Hawk in Arizona. Wilson Bull. 48: 313–314.
In Arivaipa Canyon.

———
1937. Unusual Sparrow records from Arizona. Wilson Bull. 49: 294–295.
Accounts of Swamp Sparrow, Harris's Sparrow, and Song Sparrow in northeastern Arizona.

———
1937. Notes on the birds from Graham County, Arizona. Condor 39: 254–255.
Accounts of 23 species.

———
1939. Some unusual Arizona and New Mexico bird records. Condor 41: 167–168.
Notes on 18 species in Arizona.

———
1942. Notes on some birds of southeastern Arizona. Condor 44: 222–225.
Accounts of 47 species.

———
1943. Water birds of Yuma County. Ariz. Wildlife and Sportsman 5 (7): 8.

———
1944. Notes on birds of the Yuma region. Condor 46: 19–22.
Accounts of 41 species, including Roseate Spoonbill.

———
1946. Brewster's Booby in Arizona. Auk 63: 96.
In Havasu Lake.

———
1947. Botteri's Sparrow in Arizona. Auk 64: 139–140.
Near Elgin.

———
1947. Zone-tailed Hawk breeding along Colorado River. Wilson Bull. 59: 172.

———
1948. The Starling in Arizona. Condor 50: 45.
First record, from Parker, November 16, 1946.

———
1948. Egrets nest along Colorado River. Auk 65: 603–604.
Mentions *Casmerodius albus* and *Leucophoyx thula; Phalacrocorax auritus* and *Ardea herodius* also nested in 1947.

———
1949. Recent notes from the lower Colorado River valley of Arizona and California. Condor 51: 262–265.
Brief accounts of 57 species.

———
1954. Westward extension of the ranges of the Inca Dove and Bronzed Cowbird. Condor 56: 229–230.
Includes incidental references to their Arizona distribution.

———
1961. Two year vacation. [In Letters in] Aud. Mag. 63: 131.
Photograph of a Brown Booby at Yuma. It remained there from "early September, 1958 until early October, 1960."

———
1965. A pessimistic view—the Thick-billed Parrot. Aud. Field Notes 19: 389.
Brief mention of Arizona occurrences.

———
1965. The Arizona desert. Pp. 304–311, in O. S. Pettingill, Jr., ed. The Bird Watcher's America. McGraw-Hill, New York.
A brief account of birds along the Mexican border.

———
1968. The Arizona State bird-list, 1964–67. Jour. Ariz. Acad. Sci. 5: 34–35.
Nine species are added.

——— and A. R. Phillips
1941. Bird records from southern and western Arizona. Condor 43: 108–112.
Accounts of 53 species, including the Yellow-bellied Sapsucker and White-throated Sparrow, new to the Arizona State list.

——— and A. R. Phillips
1964. An annotated check list of the species of birds in Arizona. In part 4, pp. 175–248 in C. H. Lowe, ed., The Vertebrates of Arizona, Univ. Ariz. Press, Tucson. (Part 4 is also reprinted, separately paged, as A Checklist of the Birds of Arizona, by Monson and Phillips, Univ. Ariz. Press, 1964, 74 pp.)

Moore, R. T.
1937. Four new birds from northwestern Mexico. Proc. Biol. Soc. Wash. 50: 95–102.
Atthis heloisa is regarded only as vagrant in Arizona.

Moore, R. T.
1937. New races of *Myadestes, Spizella* and *Turdus* from northwestern Mexico. Proc. Biol. Soc. Wash. 50: 201–205.
Arizona specimens of *M. townsendi* and *Spizella passerina* are listed.

———
1938. A new race of wild Turkey. Auk 55: 112–115.
Mentions specimens of *M. g. merriami* from Arizona, and includes a key to all races of the Turkey.

———
1939. A new race of *Cynanthus latirostris* from Guanajuato. Proc. Biol. Soc. Wash. 52: 57–60.
Includes measurements of Arizona specimens.

———
1939. New races of the genera *Sialia* and *Carpodacus* from Mexico. Proc. Biol. Soc. Wash. 52: 125–130.
Specimens of *S. mexicana bairdi* from Arizona are listed.

———
1941. New races of Flycatcher, Warbler and Wrens from Mexico. Proc. Biol. Soc. Wash. 54: 35–42.
Includes list of specimens of *Salpinctes obsoletus obsoletus* from Arizona.

———
1941. Three new races in the genus *Otus* from central Mexico. Proc. Biol. Soc. Wash. 54: 151–159.
Includes list of specimens of *O. asio cineraceus* from Arizona.

———
1942. Notes on *Pipilo fuscus* of Mexico and description of a new form. Proc. Biol. Soc. Wash. 55: 45–48.
Includes list of specimens of *P. f. mesoleucus* from Arizona.

———
1946. The Rufous-winged Sparrow, its legends and taxonomic status. Condor 48: 117–123.
Observations at Fresnal, Arizona.

———
1947. New species of Parrot and race of Quail from Mexico. Proc. Biol. Soc. Wash. 60: 27–28.
Specimens of *Lophortyx gambelii* from Arizona are listed.

——— and R. M. Bond
1946. Notes on *Falco sparverius* in Mexico. Condor 48: 242–244.
Mentions intergradation in the Gila River Valley of Arizona.

Morcom, G. F.
1887. Notes on the birds of southern California and southwestern Arizona. Ridgway Orn. Club Bull. 2: 36–57.
The list of 139 species includes only a few from Arizona in the vicinity of Yuma.

Neff, J. A.
1944. Banding western Doves. News from the Bird-Banders 19: 27–30.
Contains a number of Arizona records.

Newman, R. J. and G. H. Lowery, Jr.
1964. Selected quantitative data on night migration in autumn. Louisiana State Univ. Mus. Zool. Spec. Publ. 3: 1–31.
Eleven observers in Arizona participated. Migration traffic rates are given.

Nice, M. M.
1934. A Hawk census from Arizona to Massachusetts. Wilson Bull. 46: 93–95.
Six Hawks, six Owls, and three Vultures recorded in Arizona.

———
1948. Desert and mountain in southern Arizona. Aud. Bull. 67: 1–7.
A running account of the birds observed on a visit to Arizona in 1948.

Norris, J. P., Jr.
1926. A catalogue of sets of Accipitres' eggs in the collection of Joseph Parker Norris, Jr., Philadelphia, Pa., U.S.A. Oologists' Record 6 (2): 25–41.
Includes sets of *Asturina p. plagiata* and *Urubitinga a. anthracina* from Arizona.

Norris, R. A.
1958. Comparative biosystematics and life history of the Nuthatches *Sitta pygmaea* and *Sitta pusilla*. Univ. Calif. Publ. Zool. 56 (2): 119–300.
Includes specimens of *S. p. melanotis* from Arizona.

Oberholser, H. C.
1917. The migration of North American birds. Second series. I. Five Swallows. Bird-Lore 19: 320–330.
Contains migration dates of the Violet-green, Bank, and Rough-winged Swallows in Arizona.

———
1918. The migration of North American birds. Second series. II. The Scarlet and Louisiana Tanagers. Bird-Lore 20: 16–19.
Contains migration dates of the Louisiana Tanager in Arizona.

———
1918. The migration of North American birds. Second series. III. The Summer and Hepatic Tanagers, Martins, and Barn Swallows. Bird-Lore 20: 145–152.
Contains Arizona migration dates.

———
1918. The migration of North American birds. Second series. IV. The Waxwings and Phainopepla. Bird-Lore 20: 219–222.
Contains Arizona migration dates.

———
1918. The migration of North American birds. Second series. V. The Shrikes. Bird-Lore 20: 286–290.
Contains migration date of the Northern Shrike in Arizona.

———
1918. The migration of North American birds. Second series. VI. Horned Larks. Bird-Lore 20: 345–349.
Contains map of the breeding areas of the American races: four in Arizona.

Oberholser, H. C.
1919. The migration of North American birds. Second series. VIII. Ravens. Bird-Lore 21: 23–24.
Contains distribution of *C. c. sinuatus, C. c. clarionensis,* and *C. cryptoleucus* in Arizona.

———
1919. The migration of North American birds. Second series. XI. Canada Jay, Oregon Jay, Clarke's Nutcracker, and Pinon Jay. Bird-Lore 21: 354–355.
General distribution.

———
1920. The migration of North American Birds. Second series. XII. Arizona Jay, California Jay, and their allies. Bird-Lore 22: 90–91.
General distribution.

———
1920. The migration of North American birds. Second series. XIV. Cowbirds. Bird-Lore 22: 343–345.
General distribution.

———
1921. The migration of North American birds. Second series. XV. Yellow-headed Blackbird and Meadowlarks. Bird-Lore 23: 78–82.
General distribution.

———
1921. The migration of North American birds. Second series. XVII. Rusty Blackbird and Brewer Blackbird. Bird-Lore 23: 295–299.
General distribution.

———
1922. The migration of North American birds. Second series. XVIII. Red-winged Blackbirds. Bird-Lore 24: 85–88.
General distribution; *Agelaius gubernator californicus* accidental at Casa Grande.

———
1923. The migration of North American birds. Second series. XXII. Bullock's Oriole and Hooded Orioles. Bird-Lore 25: 243–244.
Contains Arizona migration dates.

———
1923. The migration of North American birds. Second series. XXIII. Scott's Oriole and Audubon's Oriole. Bird-Lore 25: 388–389.
Contains migration dates of Scott's Oriole in Arizona.

———
1924. The migration of North American birds. Second series. XXIV. Ruby-throated, Black-chinned, and Calliope Hummingbirds. Bird-Lore 26: 108–111.
Contains migration dates of the Black-chinned and Calliope Hummingbirds in Arizona.

———

1924. The migration of North American birds. Second series. XXV. Broad-billed, Rivoli, and Blue-throated Hummingbirds. Bird-Lore 26: 247–248.
Arizona migration dates.

———

1924. The migration of North American birds. Second series. XXVI. Broad-tailed, Rufous, and Allen's Hummingbirds. Bird-Lore 26: 398–399.
Contains migration dates of the Broad-tailed Hummingbird in Arizona.

———

1925. The migration of North American birds. Second series. XXVII. Costa's, Anna's, Rieffer's and Buff-bellied Hummingbirds. Bird-Lore 27: 103–104.
Contains migration dates of Costa's Hummingbird in Arizona.

———

1925. The migration of North American birds. Second series. XXVIII. Morcom's, Lucifer, Salvin's, Xantus's, and White-eared Hummingbirds. Bird-Lore 27: 326.
Distribution.

———

1926. The migration of North American birds. Second series. XXIX. The Swifts. Bird-Lore 28: 9–13.
Contains migration dates of the White-throated Swift in Arizona.

———

1926. The migration of North American birds. Second series. XXX. Chuck-will's-widow and Whip-poor-will. Bird-Lore 28: 117–120.
General distribution of the Stephen's Whip-poor-will.

———

1926. The migration of North American birds. Second series. XXXI. The Nighthawks. Bird-Lore 28: 255–261.
Contains migration dates of the Texas Nighthawk in Arizona.

———

1926. The migration of North American birds. Second series. XXXII. Parauque and Poor-will. Bird-Lore 28: 392–393.
Contains migration dates of the Poor-will in Arizona.

———

1927. The migration of North American birds. Second series. XXXIV. Red-bellied, Golden-fronted, and Gila Woodpeckers. Bird-Lore 29: 260.
General distribution.

———

1927. The migration of North American birds. Second series. XXXV. Red-headed and Lewis's Woodpeckers. Bird-Lore 29: 411–413.
General distribution.

———

1928. The migration of North American birds. Second series. XXXVI. Pileated and Ant-eating Woodpeckers. Bird-Lore 30: 112–113.
General distribution of Mearn's Woodpecker.

Oberholser, H. C.
1928. The migration of North American birds. Second series. XXXVIII. Williamson's Sapsucker and White-headed Woodpecker. Bird-Lore 30: 388.
General distribution.

———
1929. The migration of North American birds. Second series. XXXIX. Three-toed Woodpeckers. Bird-Lore 31: 110.
General distribution.

———
1929. The migration of North American birds. Second series. XLI. Texas and Nuttall's Woodpeckers. Bird-Lore 31: 398.
General distribution.

———
1930. The migration of North American birds. Second series. XLII. Hairy and Downy Woodpeckers. Bird-Lore 32: 120–123.
General distribution.

———
1930. The migration of North American birds. Second series. XLIV. The Kingfishers. Bird-Lore 32: 414–417.
General distribution.

———
1931. The migration of North American birds. Second series. XLV. The Coppery-tailed Trogon. Bird-Lore 33: 118.
General distribution.

———
1932. The migration of North American birds. Second series. XLIX. The Thick-billed Parrot. Bird-Lore 34: 263.
General distribution.

———
1932. The migration of North American birds. Second series. LI. The Elf Owl. Bird-Lore 34: 387.
General distribution.

Ohmart, R. D.
1966. Breeding record of the Cassin Sparrow *(Aimophila cassinii)* in Arizona. Condor 68: 400.
Near Tucson.

———
1968. Breeding of Botteri's Sparrow *(Aimophila botterii)* in Arizona. Condor 70: 277.
At Ophir Gulch, Santa Rita Mountains, Pima County, Arizona.

Osgood, W. H.
1903. A list of birds observed in Cochise County, Arizona. Condor 5: 128–131.
Annotated list of 72 species.

———

1903. A list of birds observed in Cochise County, Arizona. Condor 5: 149–151.
Annotated list of 51 species.

Palmer, T. S.
1918. Costa's Hummingbird—its type locality, early history, and name. Condor 20: 114–116.
Mentions nest found at Riverside, Arizona.

Peet, M. M.
1947. Violet-crowned Hummingbird in Arizona. Condor 49: 89.
From the Chiricahua Mountains.

———

1948. The Prothonotary Warbler in Arizona. Condor 50: 134.
At Cave Creek in the Chiricahua Mountains.

Pember, F. T.
1892. Collecting in the Gila Valley. Wilson Quarterly 4: 1–6.
A running account of the birds observed in April 1890 and April 1891.

———

1892. Collecting in the Gila Valley. Wilson Quarterly 4: 49–54.
A continuation of his previous paper.

Peters, J. L.
1926. N. A. races of *Falco columbarius*. Bull. Essex Co. Ornith. Club 8 (1): 20–26.
Migrates (?) through Arizona.

———

1936. Records of two species new to Arizona. Condor 38: 218.
Dryobates nuttallii at Phoenix and *Tyrannus melancholicus occidentalis* at Fort Lowell.

Peterson, R. T.
1939. Great-tailed Grackle breeding in New Mexico. Condor 41: 217.
Found nesting at Lordsburg, New Mexico, but none seen in "southeastern Arizona."

Pettingill, O. S., Jr.
1964. A convention birding trip. Aud. Mag. 66: 251.
Gives directions for finding Rufous-winged Sparrows near Tucson.

Phillips, A. R.
1932. Arizona Waxwings. Bird-Lore 34: 133.
Bohemian Waxwings observed 50 miles southwest of Tucson.

———

1933. Summer birds of a northern Arizona marsh. Condor 35: 124–125.
Accounts of 8 species.

———

1933. Further notes on the birds of the Baboquivari Mountains, Arizona. Condor 35: 228–230.
Accounts of 44 species.

Phillips, A. R.
1935. Notes from the Santa Catalina Mountains, Arizona. Condor 37: 88–89.
Accounts of 7 species.

———
1936. Bird life around Tuba City. In Kerley News, Kerley Trading Post, Tuba City, Arizona, p. 9.
Random notes; 54 species found, but only a few are listed.

———
1936. Status of the Marbled Godwit in Arizona. Condor 38: 215–216.
Regarded as a common migrant on Mormon Lake.

———
1940. Two new breeding birds for the United States. Auk 57: 117–118.
Tyrannus melancholicus occidentalis and *Cassidix mexicanus nelsoni* from Tucson.

———
1942. Notes on the migrations of the Elf and Flammulated Screech Owls. Wilson Bull. 54: 132–137.
Both are considered to be migratory in Arizona. No winter records.

———
1944. Some differences between the Wright's and Gray Flycatchers. Auk 61: 293–294.
Includes notes on their ranges in Arizona.

———
1944. Status of Cassin's Sparrow in Arizona. Auk 61: 409–412.

———
1946. [On R. T. Moore's "legends" concerning the Rufous-winged Sparrow.] [In] Notes and News, Condor 48: 249–250.
A criticism of Moore's identifications.

———
1947. Records of occurrence of some southwestern birds. Condor 49: 121–123.
Some of the records pertain to New Mexico.

———
1947. Bird life of the San Francisco Mountains. No. 5: Hawks and Owls. Plateau 20: 17–22.

———
1947. Status of the Anna Hummingbird in southern Arizona. Wilson Bull. 59: 111–113.
An "autumn visitant."

———
1949. Nesting of the Rose-throated Becard in Arizona. Condor 51: 137–139.
In ". . . the Santa Cruz River drainage."

———
1950. The San Blas Jay in Arizona. Condor 52: 86.
Near Tucson.

———
1951. Complexities of migration: a review with original data from Arizona. Wilson Bull. 63: 129–136.
An important critical paper.

———
1954. Western records of *Chaetura vauxi tamaulipensis*. Wilson Bull. 66: 72–73.
At Fort Huachuca.

———
1955. The history of Say's Phoebe at Flagstaff. Plateau 28: 25–28.

[Phillips, A. R.]
1956. [Duck Hawks take over Prairie Falcon nesting sites in Arizona.] [In] Notes and News [of Cooper Society Meetings.] Condor 58: 240.

Phillips, A. R.
1956. The migrations of birds in northern Arizona. Plateau 29: 31–35.

———
1957. The Vaux Swift in western Mexico. Condor 59: 140–141.
Incidental mention of an Arizona specimen.

———
1958. Las peculiaridades del Sastrecito (*Psaltriparus* familia Paridae) y su incubacion. Univ. Mex., Anales Inst. de Biol. 29: 355–360.
Includes discussion of Arizona specimens.

———
1959. La acrecencia de errores acerca de la ornitologia de México, con notas sobre *Myiarchus*. Univ. Méx., Anales del Inst. de Biol. 30: 349–368.
Includes comments on Arizona specimens.

———
1959. Souces of error in banding and homing studies. Bird-Banding 30: 229–231.
Westward migration into Arizona is mentioned.

———
1961. Notas sistematicas sobre aves mexicanas. I. Univ. Méx., Anales del Inst. de Biol. 32: 333–381.
Includes comments on some Arizona species.

———
1961. Emigraciones y distribucion de aves terrestres en Mexico. Rev. Soc. Mex. Hist. Nat. 22: 295–311.
Includes brief comments on migration of several birds through southwestern Arizona.

———
1962. Notas sistematicas sobre aves mexicanas. II. Univ. Méx., Anales del Inst. de Biol. 33: 331–372.
Includes comments on some Arizona species.

Philips, A. R.

1966. Further systematic notes on Mexican birds. Bull. British Ornith. Club 86: 86–131.

Arizona specimens mentioned: *Peucedramus taeniatus* and *Vireo huttoni stephensi.*

———

1968. The instability of the distribution of land birds in the southwest. [In] A. H. Schroeder. Collected papers in honor of Lyndon Lane Hargrave. Papers Archeol. Soc. New Mex. 1: 129–162.

Much criticism of past and present ornithological studies; 85 titles in bibliography.

——— and D. Amadon

1952. Some birds of northwestern Sonora, Mexico. Condor 54: 163–168.

Includes brief comments on some Arizona specimens.

——— and R. W. Dickerman

1957. Notes on the Song Sparrows of the Mexican plateau. Auk 74: 376–382.

They know of "no valid record of *merrilli* in eastern or central Arizona."

——— and W. M. Pulich

1948. Nesting birds of the Ajo Mountains region, Arizona. Condor 50: 271–272.

Accounts of 11 species.

——— and J. D. Webster

1961. Grace's Warbler in Mexico. Auk 78: 551–553.

Measurements of specimens from Arizona are listed.

Phillips, P.

1937. Dusky Grouse in the Chuskai Mountains of northeastern Arizona and northwestern New Mexico. Auk 54: 203–204.

A pair observed 8 miles southeast of Lukachukai, Apache County, Arizona.

[Pierce, W. M.]

1934. [Report of Black Vultures near Sells, Arizona.] [In] Minutes of Cooper Club Meetings, Condor 36: 124.

Pitelka, F. A.

1941. Distribution of birds in relation to major biotic communities. Amer. Midland Nat. 25: 113–137.

Some Arizona species are included in the examples.

———

1948. Notes on the distribution and taxonomy of Mexican game birds. Condor 50: 113–123.

Includes measurements of *Lophortyx gambelii gambelii* from Arizona and mentions that *Zenaida asiatica* winters in southern Arizona.

———

1950. Geographic variation and the species problem in the shore-bird genus *Limnodromus.* Univ. Calif. Publ. Zool. 50: 1–108.

A specimen of *L. scolopaceus* from Mimbres, Cochise County, Arizona, is listed.

Plummer, C. M.
1934. Sundry observations. Bird-Lore 36: 297.
At Phoenix.

Poling, O. C.
1890. The presence of McCown's and the Chestnut-collared Longspur in southern Arizona, near the Mexican border. Ornith. and Ool. 15 (5): 71.
At Fort Huachuca during February and March.

——
1891. Groove-billed Ani *(Crotophaga sulcirostris)* in Arizona. Auk 8: 313–314.
One collected May 1888 in the Huachuca Mountains.

Poor, H. H.
1946. Western habitats. Aud. Mag. 48: 207–211.
A running account of the birds from the mountain peaks to the desert in Arizona.

Price, W. W.
1899. Some winter birds of the lower Colorado Valley. Bull. Cooper Ornith. Club 1: 89–93.
Ninety-one species recorded between Yuma and the mouth of the Colorado River.

Pugh, E. A.
1954. The status of birds in the Mount Elden area. Plateau 26: 117–123.
Accounts of 96 species.

Pulich, W.
1947. Another record of the Wood Duck in Arizona. Condor 49: 131.
At Phoenix.

Pulich, W. M.
1950. Second record of the Ross Goose from Arizona. Condor 52: 90.
Near Topock.

——
1952. The Arizona Crested Flycatcher in Nevada. Condor 54: 169–170.
Includes notes on Arizona distribution.

—— and A. R. Phillips
1951. Autumn bird notes from the Charleston Mountains, Nevada. Condor 53: 205–206.
Incidental remarks on Arizona specimens.

—— and A. R. Phillips
1953. A possible desert flight line of the American Redstart. Condor 55: 99–100.
Along the lower Colorado River.

Quaintance, C. W.
1935. Water Ouzel nests on Black River, Arizona. Condor 37: 43.

——
1935. Notes from central eastern Arizona. Condor 37: 83–84.
Accounts of 7 species.

[Radke, E. L., ed.]
1966. [Information sought on tagged Cowbirds.] Western Bird Bander 41: 22.
Mentions that 434 Brown-headed Cowbirds were banded in Phoenix in 1966.

———
1966. [Cal Royall desires information on Brown-headed Cowbirds which he banded in Phoenix, Arizona.] Western Bird Bander 41: 22.

———
1966. [Information sought on Verdins.] Western Bird Bander 41: 28.
Includes mention that Walter Taylor has banded to date about 90 Verdins in his study of their breeding biology in the Tempe, Arizona, region.

Raitt, R. J. and R. D. Ohmart
1966. Annual cycle of reproduction and molt in Gambel Quail of the Rio Grande Valley, southern New Mexico. Condor 68: 541–561.
Includes brief comments (pp. 556–557) on Gambel Quail in dry years in Arizona.

Rand, A. L.
1942. Results of the Archbold expeditions. No. 44. Some notes on bird behavior. Bull. Amer. Mus. Nat. Hist. 79: 517–524.
Gambel Quail, Roadrunner, and Great Horned Owl at Tucson, Arizona.

Reis, C. O.
1938. Long-crested Jay in southeastern California. Condor 40: 44.
Near Blythe, along the Colorado River.

Rhoads, S. N.
1892. The birds of southeastern Texas and southern Arizona observed during May, June and July, 1891. Proc. Acad. Nat. Sci. Phila.: 98–126.
Accounts of 126 species from Arizona.

———
1893. The *Vireo huttoni* group, with description of a new race from Vancouver Island. Auk 10: 238–241.
Specimens of *V. h. stephensi* from Arizona examined.

Ridgway, R.
1869. Notices of certain obscurely known species of American birds. (Based on specimens in the museum of the Smithsonian Institution.) Proc. Acad. Nat. Sci. Phila.: 125–135.
Includes comments on *Piranga cooperi* at Ft. Mohave.

———
1872. On the occurrence of *Setophaga picta* in Arizona. Amer. Nat. 6: 436.
Near Tucson.

———
1873. The birds of Colorado. Bull. Essex Inst. 5: 174–195.
Mentions, on p. 178, *Coccygus Americanus* breeding at Tucson, Arizona.

———
1874. Two rare Owls from Arizona. Amer. Nat. 8: 239–240.
Syrnium occidentale and *Micrathene whitneyi,* both from vicinity of Tucson.

———

1875. On *Nisus cooperi* (Bonaparte) and *N. gundlachi* (Lawrence). Proc. Acad. Nat. Sci. Phila.: 78–88.
Includes measurements of a specimen of *N. cooperi* from Arizona.

———

1875. On the Buteonine subgenus *Craxirex* Gould. Proc. Acad. Nat. Sci. Phila.: 89–119.
Includes remarks on an Arizona specimen of *Buteo swainsoni.*

———

1876. Studies of the American Falconidae. U.S. Geol. and Geog. Surv. Terr. Bull. vol. II, no. 2: 91–182.
Lists Arizona specimens in the U.S. National Museum.

———

1878. Studies of the American Herodiones. Part I. U.S. Geol. and Geog. Surv. Terr. Bull. IV, Art. IX: 219–251.
Includes synonymy of *Ardea herodias.*

———

1878. A review of the American species of the genus *Scops,* Savigny. Proc. U.S. Nat. Mus. 1: 85–117.
Lists Arizona specimens of *S. asio, S. trichopsis* and *S. flammeolus.*

———

1880. Catalogue of Trochilidae in the collection of the United States National Museum. Proc. U.S. Nat. Mus. 3: 308–320.
Lists nine species from Arizona.

———

1881. A review of the genus *Centurus,* Swainson. Proc. U.S. Nat. Mus. 4: 93–119.
Includes *C. uropygialis,* with synonymy, description, and distribution.

———

1882. List of additions to the catalogue of North American birds. Bull. Nuttall Orn. Club 7: 257–258.

———

1884. *Ortyx virginianus* not in Arizona. Forest and Stream 22: 124.
Either *Cyrtonyx massena* or *Ortyx graysoni.*

———

1884. A review of the American Crossbills *(Loxia)* of the *L. curvirostra* type. Proc. Biol. Soc. Wash. 2: 101–107.
Includes a discussion of Arizona specimens.

———

1887. The Coppery-tailed Trogon *(Trogon ambiguus)* breeding in southern Arizona. Auk 4: 161–162.
In the Huachuca Mountains.

———

1887. *Trogon ambiguus* breeding in Arizona. Proc. U.S. Nat. Mus. 10: 147.
In the Huachuca Mountains.

Robbins, C. S.
1948. [Breeding distribution of the Vesper Sparrow.] Aud. Field Notes 2: 147.
Map includes localities in Arizona.

———
1948. [Breeding distribution of the Whip-poor-will.] Aud. Field Notes 2: 192.
Map includes localities in Arizona.

———
1949. Migration of the Redhead, pp. 25–28, with 4 maps, in Migration of some North American Waterfowl. U.S. Dept. Int. Fish and Wildl. Serv. Spec. Sci. Rep. (Wildl.) no. 1.

———
1949. Distribution of North American birds. The breeding distribution of the Virginia Rail. Aud. Field Notes 3: 238–239.
Map shows one record in Arizona.

———
1949. Distribution of North American birds. Aud. Field Notes 3: 262–264.
Map of winter distribution of Virginia Rail includes Arizona.

———
1950. Distribution of North American birds. Winter distribution of the Sora. Aud. Field Notes 4: 40.
Map shows three Arizona localities.

——— and W. T. Van Velzen
1969. The breeding bird survey 1967 and 1968. U.S. Dept. Int. Bur. Sport Fisheries and Wildl. Spec. Sci. Rep. (Wildl.) 124: 1–107.
Of Arizona interest are Mourning Dove, Red-shafted Flicker, Western Kingbird, Horned Lark, House Finch, and Rufous-sided Towhee.

Roberts, T. S.
1935. Birds seen in the Salt and Gila Rivers region, Arizona in March and April, 1933. Ariz. Wild Life 6 (9): 1, 3.
A list of 72 species.

Roest, A. I.
1957. Notes on the American Sparrow Hawk. Auk 74: 1–19.
Sex ratios of Arizona specimens are given.

Root, R. B.
1962. Comments on the status of some western specimens of the American Redstart. Condor 64: 76–77.
Discusses Arizona specimens and doubts that the species breeds in the state.

Royall, W. C., Jr.
1968. Cowbirds at a Phoenix, Arizona, cattle feedlot. Western Bird Bander 43: 41–42.
Records of Brown-headed and Bronzed Cowbirds.

Ruos, J. L. and D. MacDonald
1968. Mourning Dove status report, 1967. U.S. Dept. Int. Bur. Sport Fisheries and Wildl. Spec. Sci. Rep. (Wildl.) 121: 1–23.

——— and D. MacDonald
1970. Mourning Dove status report, 1968. U.S. Dept. Int. Bur. Sport Fisheries and Wildl. Spec. Sci. Rep. (Wildl.) 129: 1–38.

——— and R. E. Tomlinson
1968. Mourning Dove status report, 1966. U.S. Dept. Int. Bur. Sport Fisheries and Wildl. Spec. Sci. Rep. (Wildl.) 115: 1–49.

Russell, H. N., Jr.
1935. Report of field work in ornithology. Rainbow Bridge Monument Valley expedition. N.Y.A. Project 3968-Y-1, Berkeley. 20 pp.
Accounts of 86 species and subspecies.

Ryder, R. A.
1963. Migration and population dynamics of American Coots in western North America. Proc. 13th Int. Ornith. Cong. Ithaca, 17–24, June 1962, 1: 441–453.
Includes maps of distribution and recoveries—some in Arizona.

———
1967. Distribution, migration and mortality of the White-faced Ibis (*Plegadis chihi*) in North America. Bird-Banding 38: 257–277.
Includes Arizona distribution with one recovery of a banded bird.

Salt, W. R.
1931. Banding operations at Rosebud, Alberta during 1930–31. News from the Bird-Banders 6 (3): 6–7.
A banded Pigeon Hawk was shot in the Gila Valley, Arizona.

San Miguel, M.
1968. Commentary on the 1967 annual report. Western Bird Bander 43: 37.
Includes banding summary, 1961 to 1967; Arizona and New Mexico data are combined.

[Sargent, G., ed.]
1936. [Say's Phoebe banded at Casa Grande Indian Ruins, Arizona.] [In] Digest of the minutes of the Los Angeles Chapter. News from the Bird-Banders 11: 35–37.

Saunders, G. B.
1951. A new White-winged Dove from Guatemala. Proc. Biol. Soc. Wash. 64: 83–87.
Lists specimens of *Zenaida asiatica mearnsi* from Arizona.

———
1968. Seven new White-winged Doves from Mexico, Central America, and southwestern United States. N.A. Fauna 65: 1–30.
None are from Arizona, but comparisons are made with Arizona specimens.

Schaefer, O. F.
1917. Occurrence of the Red-breasted Nuthatch in Arizona. Condor 19: 103.
Observed 40 miles south of Winslow.

[Scheffler, W. J.]
1941. [Mearns' Quail numerous in Arizona.] Condor 43: 208.

Schellbach, L.
1933. [Brewer's Sparrow and Killdeer at the Grand Canyon.] Grand Canyon Nature Notes 8: 215–216.

Sclater, P. L.
1862. Catalogue of a collection of American birds belonging to Philip Lutley Sclater, M.A., Ph.D., F.R.S. N. Trubner and Co., London. i–xvi, 1–368, pls. 20.
Lists, on p. 230, *Empidonax obscurus* from Fort Yuma.

Scott, W. E. D.
1885. Winter mountain notes from southern Arizona. Auk 2: 172–174.
Running account of the birds observed in the Santa Catalina Mountains from November 26–29, 1884.

———
1886. On the avi-fauna of Pinal County, with remarks on some birds of Pima and Gila Counties, Arizona. With annotations by J. A. Allen. Auk 3: 249–258; 383–389; 421–432. Auk 4: 16–24; 196–205. Auk 5: 29–36; 159–168.

Searl, C.
1930. A new bird record. Grand Canyon Nature Notes 4: 80.
Sight record of Black-billed Magpies at Bright Angel Point, North Rim, Grand Canyon.

———
1931. Early November bird records. Grand Canyon Nature Notes 6: 3–6.
A list of 16 species at Yavapai Point, Grand Canyon.

Selander, R. K.
1954. A systematic review of the Booming Nighthawks of western North America. Condor 56: 57–82.
Breeding specimens of *C. m. henryi* from Arizona are listed; *C. m. hesperis* is a migrant.

——— and D. R. Giller
1961. Analysis of sympatry of Great-tailed and Boat-tailed Grackles. Condor 63: 29–86.
Map of distribution of Arizona races of *Cassidix mexicanus* is included.

——— and D. R. Giller
1963. Species limits in the Woodpecker genus *Centurus (Aves)*. Bull. Amer. Mus. Nat. Hist. 124, art. 6: 213–274.
Includes data on the Gila Woodpecker in Arizona.

——— and M. A. del Toro
1955. A new race of Booming Nighthawk from southern Mexico. Condor 57: 144–147.
Lists specimens of *C. m. henryi* from Arizona.

Seubert, J. L.
1963. Research on methods of trapping the Red-winged Blackbird *(Agelaius phoeniceus)*. Angewandte Ornith. 1: 163–170.
Includes a map (p. 167) of the location of the 1961–1962 winter Blackbird and Starling roosts, of which five are pictured in Arizona.

Sheppard, J. M.
1968. Berylline and Violet-crowned Hummingbirds in Arizona. Auk 85: 329.
In Ramsey Canyon, Huachuca Mountains, Arizona.

Sibley, C. G.
1950. Species formation in the Red-eyed Towhees of Mexico. Univ. Calif. Publ. Zool. 50: 109–194.
Specimens of *Pipilo erythrophthalmus montanus* from Arizona were examined.

——— and L. L. Short, Jr.
1959. Hybridization in the Buntings *(Passerina)* of the Great Plains. Auk 76: 443–463.
Map of breeding range of *P. amoena.*

——— and L. L. Short, Jr.
1964. Hybridization in the Orioles of the Great Plains. Condor 66: 130–150.
Specimens from Arizona examined.

Simpson, J. M. and J. R. Werner
1958. Some recent bird records from the Salt River Valley, central Arizona. Condor 60: 68–70.
Accounts of 23 species.

Sloanaker, J. L.
1912. Two new Arizona records. Condor 14: 154.
Grus canadensis and *Clangula clangula americana* at Tucson.

———
1913. Bird notes from the south-west. Wilson Bull. 25: 187–199.
Contains a list of 45 species observed in the vicinity of Tucson.

Smiley, D. C.
1937. Water birds of the Boulder Dam region. Condor 39: 115–119.
A census of 22 species.

Smith, A. G.
1949. Migration of the Baldpate, pp. 11–12, with map, in Migration of some North American Waterfowl, U.S. Dept. Int. Fish and Wildl. Serv. Spec. Sci. Rep. (Wildl.) no. 1.

———
1949. Migration of the Ruddy Duck, pp. 45–46, with map, in Migration of some North American Waterfowl, U.S. Dept. Int. Fish and Wildl. Serv. Spec. Sci. Rep. (Wildl.) no. 1.

Smith, A. P.
1907. The Thick-billed Parrot in Arizona. Condor 9: 104.
Observed in the Chiricahua Mountains.

———
1907. Summer notes from an Arizona camp. Condor 9: 196–197.
Accounts of 24 species of birds observed at Benson and in the Whetstone Mountains.

———
1908. Is the Mountain Bluebird resident at high altitudes? Condor 10: 50.
At Flagstaff in February and March, 1907.

Smith, A. P.
1908. Some data and records from the Whetstone Mountains, Arizona. Condor 10: 75–78.
Brief accounts of birds observed in the summer months.

———
1918. Whip-poor-will in New Mexico in March. Condor 20: 91–92.
Also observed in the Chiricahua Mountains in May.

Stannard, C.
1939. Mourning Dove returns. News from the Bird-Banders 14: 36.
A Mourning Dove banded at Phoenix was killed in Mexico.

———
1941. Regularity in time of return of individuals. News from the Bird-Banders 16: 26–27.
Concerns Gambel Sparrows at Phoenix.

———
1944. Trapping Doves and Pigeons. News from the Bird-Banders 19: 32–33.
Experiences at Phoenix.

Starck, F. L.
1947. Winter birds of Arizona. Jack-Pine Warbler 25: 12–16.
A running account of the birds observed in the Tucson vicinity in January and part of February 1942.

——— and R. Starck
1947. Birds observed in the vicinity of Tucson, Arizona, in January, 1942. Jack-Pine Warbler 25: 16–17.
A list of 89 species.

Stein, L.
1934. Statistical report of bird banding. Grand Canyon Nature Notes 9: 274–276.
Summaries of 13 species banded in Grand Canyon National Park.

Stensrude, C.
1965. Observations on a pair of Gray Hawks in southern Arizona. Condor 67: 319–321.
At Rillito Creek, Pima County.

Stephens, F. [= F. Stevens in byline.]
1878. Notes on a few birds observed in New Mexico and Arizona in 1876. Bull. Nuttall Ornith. Club 3: 92–94.
Twenty species listed, but localities not always clearly stated.

Stephens, F.
1885. Notes on an ornithological trip in Arizona and Sonora. Auk 2: 225–231.
Running account of the birds observed on a trip from Tucson to the Gulf of California.

———
1903. Bird notes from eastern California and western Arizona. Condor 5: 75–78.
The Arizona observations are from Needles to the Hualapai Mountains.

———
1903. Bird notes from eastern California and western Arizona. Condor 5: 100–105.
Annotated list, in Arizona from Needles to the Hualapai Mountains.

———
1913. Early nesting of the Band-tailed Pigeon. Condor 15: 129.
The nest alluded to on page 124 of Bendire's Life Histories of North American birds was in California, not Arizona.

———
1914. Arizona records. Condor 16: 259.
Notes on *Columba fasciata, Ammodramus bairdi,* and *Passerella iliaca schistacea.*

Stevenson, J. O.
1936. Bird notes from the Hualpai Mountains, Arizona. Condor 38: 244–245.
Records of 14 species.

———
1938. The Timberline Sparrow in Arizona. Condor 40: 86–87.
One collected near Springerville.

Stewart, R. E. and J. W. Aldrich
1956. Distinction of maritime and prairie populations of Blue-winged Teal. Proc. Biol. Soc. Wash. 69: 29–34.
Includes list of specimens of *Anas discors discors* from Arizona.

———, A. D. Geis, and C. D. Evans
1958. Distribution of populations and hunting kill of the Canvasback. Jour. Wild. Mgmt. 22: 333–370.
Includes maps of Arizona distribution.

Still, D. A.
1919. Observations taken at Madera Canyon, in the Santa Rita Mountain, between June 1st and June 14th, 1919. Ool. 36: 191.
Nine species listed.

Stone, W.
1890. Catalogue of the Owls in the collection of the Academy of Natural Sciences of Philadelphia. Proc. Acad. Nat. Sci. Phila.: 124–131.
Listed from Arizona: *Micropallas whitneyi.*

———
1891. Catalogue of the Corvidae, Paradiseidae, and Oriolidae in the collection of the Academy of Natural Sciences of Philadelphia. Proc. Acad. Nat. Sci. Phila.: 441–450.
Listed from Arizona: *Aphelocoma sieberii arizonae* and *Cyanocitta stelleri macrolopha.*

———
1897. The genus *Sturnella.* Proc. Acad. Nat. Sci. Phila.: 146–152.
Discusses specimens from Arizona.

Stone, W.
1905. On a collection of birds and mammals from the Colorado delta, Lower California. With field notes by Samuel N. Rhoads. Proc. Acad. Nat. Sci. Phila.: 676–690.
Includes observations in the Yuma region.

Stoner, E. A.
1962. Some returns and recoveries during 1961. Western Bird Bander 37: 32–34.
A Pintail banded at Ruby Lake, Nevada, was recovered at Crescent Lake, Arizona.

Stophlet, J. J.
1959. Nesting concentration of Long-eared Owls in Cochise County, Arizona. Wilson Bull. 71: 97–99.
Near Tombstone.

Storer, R. W.
1951. Variation in the Painted Bunting *(Passerina ciris)*, with special reference to wintering populations. Occas. Papers Mus. Zool. Univ. Mich. 532: 1–12.
Seven Arizona specimens examined.

———
1952. Variation in the resident Sharp-shinned Hawks of Mexico. Condor 54: 283–289.
Mentions intergrades in Arizona.

Stoudt, J. H.
1949. Southward migration of the Shoveller, pp. 24, with 2 maps. In Migration of some North American Waterfowl. U.S. Dept. Int. Fish and Wildl. Serv. Spec. Sci. Rept. (Wildl.) no. 1.

Strong, R. M.
1901. A quantitative study of variation in the smaller North-American Shrikes. Amer. Nat. 35: 271–298.
Included in the study were 29 specimens of Shrikes from Arizona.

Sturdevant, G. E.
1926. Water Ouzel *(Cinclus mexicanus)*. Grand Canyon Nature Notes 1 (2): 4–5.

———
1926. Long-crested Jay *(Cyanocitta stelleri diademata)*. Grand Canyon Nature Notes 1 (3): 4–5.

———
1926. The Turkey Buzzard *(Cathartes aura)*. Grand Canyon Nature Notes 1 (7): 1–2.

———
1926. A winter resident. Grand Canyon Nature Notes 1 (11): 1.
Robins.

Sturdevant, G. E., Mrs.
1927. Crossbill. Grand Canyon Nature Notes 2 (4): 1–2.

———

1927. Crossbill winters at Grand Canyon. Grand Canyon Nature Notes 2 (7): 3.

———

1928. Bird baths. Grand Canyon Nature Notes 3 (4): 4–5.

Summers-Smith, D.

1963. The House Sparrow. Collins, St. Jame's Place, London: xvi + 269 pp.
Includes summary of its introduction and its spread in the United States.

Sutton, G. M.

1941. First impressions of the bird life of southern Arizona. Proc. Wilson Ornith. Club. Wilson Bull. 53: 64.
A summary of his paper given before the 26th meeting in Minneapolis.

———

1943. Records from the Tucson region of Arizona. Auk 60: 345–350.
Accounts of 30 species.

———

1948. The Curve-billed Thrasher in Oklahoma. Condor 50: 40–43.
Includes measurements of Arizona specimens of *T. c. celsum.*

——— and O. S. Pettingill, Jr.

1943. Birds of Linares and Galeana, Nuevo Leon, Mexico. Occas. Papers Mus. Zool. Louisiana State Univ. 16: 273–291.
Incidental mention of Arizona specimens of *Callipepla squamata* and *Tachycineta thalassina.*

——— and A. R. Phillips

1942. June bird life of the Papago Indian Reservation, Arizona. Condor 44: 57–65.
Accounts of 89 species.

———, A. R. Phillips, and L. L. Hargrave

1941. Probable breeding of the Beautiful Bunting in the United States. Auk 58: 265–266.
West slope of the Baboquivari Mountains.

——— and J. Van Tyne

1935. A new Red-tailed Hawk from Texas. Occas. Papers Mus. Zool. Univ. Mich. 321: 1–6.
Arizona specimens of *B. j. calurus* are discussed.

Sutton, M.

1954. Bird survey of the Verde Valley. Plateau 27: 9–17.
A list of 179 birds, including subspecies.

Swarth, H. S.

1904. Birds of the Huachuca Mountains, Arizona. Pac. Coast Avi. 4: 1–70.
Annotated list of 196 species.

———

1904. The status of the southern California Cactus Wren. Condor 6: 17–19.
Heleodytes brunneicapillus couesi occurs from the Rio Grande to the Pacific Coast.

Swarth, H. S.
1905. Summer birds of the Papago Indian Reservation and of the Santa Rita Mountains, Arizona. Condor 7: 22–28; 47–50; 77–81.
Annotated separate lists from the two localities.

———
1908. Some fall migration notes from Arizona. Condor 10: 107–116.
Annotated list of 109 species of birds observed in the Rincon and Huachuca mountains, Arizona.

———
1909. Distribution and molt of the Mearn's Quail. Condor 11: 39–43.

———
1910. Miscellaneous records from southern California and Arizona. Condor 12: 107–110.
Atthis morcomi and *Cynanthus latirostris* from Arizona.

———
1914. A distributional list of the birds of Arizona. Pac. Coast Avi. 10: 1–133.
Contains 362 species and subspecies, a hypothetical list of 24 species, and a bibliography up to 1914.

———
1914. A study of the status of certain island forms of the genus *Salpinctes*. Condor 16: 211–217.
Specimens of *S. o. obsoletus* from Arizona examined.

———
1916. The Pacific Coast races of the Bewick Wren. Proc. Calif. Acad. Sci. 4th ser. 6 (4): 53–85.
Contains data on *T. b. eremophilus* of Arizona.

———
1917. A revision of the Marsh Wrens of California. Auk 34: 308–318.
Gives winter records of *Telmatodytes palustris plesius* from the lower Colorado River.

———
1918. Notes on some birds from central Arizona. Condor 20: 20–24.
Account of species observed along the Apache Trail between Phoenix and Globe. *Loxia curvirostra bendirei* and *Passerina cyanea* are new to the state list.

———
1918. The Pacific Coast Jays of the genus *Aphelocoma*. Univ. Calif. Publ. Zool. 17: 405–422.
Specimens of *A. woodhousei* from Arizona were examined.

———
1920. Birds of the Papago Saguaro National Monument and the neighboring region, Arizona. U.S. Dept. Int. Nat. Park Serv.: 1–63.

———
1920. Revision of the avian genus *Passerella*, with special reference to the distribution and migration of the races in California. Univ. Calif. Publ. Zool. 21: 75–224.
Arizona specimens are listed.

———
1921. *Bubo virginianus occidentalis* in California. Condor 23: 136.
Discusses specimens of *B. v. pallescens* from Arizona.

———
1922. Birds and mammals of the Stikine River region of northern British Columbia and southeastern Alaska. Univ. Calif. Publ. Zool. 24: 125–314.
Includes discussion of specimens of *Junco oreganus shufeldti* from Arizona.

———
1924. Birds and mammals of the Skeena River region of northern British Columbia. Univ. Calif. Publ. Zool. 24: 315–394.
Includes measurements of *Dryobates pubescens leucurus* from Arizona.

———
1924. Fall migration notes from the San Francisco Mountain region, Arizona. Condor 26: 183–190.
Accounts of 91 species.

———
1926. Report on a collection of birds and mammals from the Atlin region, northern British Columbia. Univ. Calif. Publ. Zool. 30: 51–162.
Arizona is included in the distributional map of the genus *Dendragapus* and of *Zonotrichia leucophrys*.

[Swarth, H. S.]
1929. [Reports Black Vulture near Tucson, Arizona.] [In] Minutes of Cooper Club Meetings, Condor 31: 83.

Swarth, H. S.
1929. The faunal areas of southern Arizona: a study in animal distribution. Proc. Calif. Acad. Sci. 4th ser. 18: 267–383.
Accounts of 164 species and subspecies of birds.

[Swarth, H. S.]
1933. [Black Vulture near Tucson, and between Tucson and Yuma.] [In] Minutes of Cooper Club Meetings, Condor 35: 47.

Swarth, H. S.
1933. Relationships of Coues and Olive-sided Flycatchers. Condor 35: 200–201.
Notes on habits of the Coues' Flycatcher in Arizona; the two species are believed to be congeneric.

———
1935. Systematic status of some northwestern birds. Condor 37: 199–204.
Discusses Arizona specimens of *Tringa solitaria* and *Vermivora celata*.

———
1936. Savannah Sparrow migration routes in the northwest. Condor 38: 30–32.
Map shows *P. s. alaudinus* migration route across Arizona.

——— and W. W. Swarth
1920. Some winter birds at the Grand Canyon, Arizona. Condor 22: 79–80.
Observations on 20 species in December 1919.

Swenk, M. H.
1930. The Crown Sparrows *(Zonotrichia)* of the Middle West. Wilson Bull. 42: 81–95.
Includes map of distribution of *Z. leucophrys* in Arizona.

———
1936. A study of the distribution, migration and hybridism of the Rose-breasted and Rocky Mountain Black-headed Grosbeaks in the Missouri Valley region. Nebraska Bird Rev. 4: 27–40.
Contains a map of breeding ranges.

——— and O. A. Stevens
1929. Harris's Sparrow and the study of it by trapping. Wilson Bull. 41: 129–177.
One Arizona record mentioned.

——— and J. B. Swenk
1928. Some impressions of the commoner winter birds of southern Arizona. Wilson Bull. 40: 17–29.
Observations, chiefly from the Tucson vicinity, from October 19, 1926, to April 27, 1927.

Swinburne, J.
1888. Breeding of the Evening Grosbeak *(Coccothraustes vespertina)* in the White Mountains of Arizona. Auk 5: 113–114.

———
1888. Occurrence of the Chestnut-collared Longspur *(Calcarius ornatus)* and also of Maccown's Longspur *(Rhynchophanes maccownii)* in Apache Co., Arizona. Auk 5: 321–322.

Tanner, J. T. and J. W. Hardy
1958. Summer birds of the Chiricahua Mountains, Arizona. Amer. Mus. Novit. 1866: 1–11.
Accounts of 114 species.

Tanner, V. M.
1936. The Western Mockingbird in Utah. Proc. Utah Acad. Sci. Arts and Letters, 13: 185–187.
Occurs "south through San Juan County to Arizona. . . ."

Taylor, W. P. and A. J. Duvall
1951. The Lucifer Hummingbird in the United States. Condor 53: 202–203.
Remarks on an Arizona specimen.

——— and C. T. Vorhies
1933. The Black Vulture in Arizona. Condor 35: 205–206.
A summary of observations in the Santa Cruz valley and westward.

——— and C. T. Vorhies
1936. The Clark Nutcracker in extreme southwestern Arizona. Condor 38: 42.
One observed at Bates Well and another near Tinajas Altas.

Thompson, B. H.
1933. Toroweap the new Grand Canyon National Monument. Grand Canyon Nature Notes 8: 162–169.
Brief list of the birds in the monument.

Thornburg, F.
1956. Rose-breasted Grosbeak in Arizona. Condor 58: 447.
Observed at Madera Canyon, Santa Rita Mountains, and at Patagonia.

Tinkham, E. R.
1949. Notes on nest-building of the Vermilion Flycatcher. Condor 51: 230–231.
At the San Bernardino Ranch, "near Tucson." (?)

Todd, W. E. C.
1913. A revision of the genus *Chaemepelia*. Ann. Carnegie Mus. 8: 507–603.
Description, synonymy, and Arizona references of *Chaemepelia passerina pallescens*.

Tomlinson, R. E.
1965. Mourning Dove status report, 1965. U.S. Dept. Int. Bur. Sport Fisheries and Wildl. Spec. Sci. Rep. (Wildl.) 91: 1–37.

Torrey, B.
1907. The Vermilion Flycatcher at Santa Barbara. Condor 9: 109.
Also observed at Tucson, Arizona.

Townsend, C. W.
1925. Winter birds seen at the Grand Canyon, Arizona. Condor 27: 177–178.
Records of 26 species observed in December 1919.

Truett, J. C.
1967. Movements of immature Mourning Doves into Mexico. Pp. 18–21 in Proc. 6th Annual Meeting of the Wildlife Society, New Mexico-Arizona Section, at Farmington, New Mexico, Feb. 3–4, 1967.
Older immatures move south in midsummer.

Tucson Audubon Society
1964. Birds of southeastern Arizona. Tucson Audubon Soc. 32 pp.
A list of 260 "regular" species with an appendix of 82 "irregular or accidental" species. Habitat and seasonal occurrence are given.

Uren, L. S.
1965. Audubon stamp bird moves to the U.S. Aud. Mag. 67: 101–102.
Calocitta formosa colliei recorded from Douglas, Arizona.

van den Akker, J. B. and V. T. Wilson
1949. Twenty years of bird banding at Bear River Migratory Bird Refuge, Utah. Jour. Wildl. Mgmt. 13: 359–376.
Returns from Arizona are included.

van Rossem, A. J.
1926. The California forms of *Agelaius phoeniceus* (Linnaeus). Condor 28: 215–230.
Discussion includes *A. p. sonoriensis* in Arizona.

———
1928. A northern race of the Mountain Chickadee. Auk 45: 104–105.
Measurements of *P. g. gambeli* from Arizona are given.

van Rossem, A. J.
1930. Four new birds from northwestern Mexico. Trans. San Diego Soc. Nat. Hist. 6: 213–226.
Comments on Arizona specimens of *Passerculus sandwichensis* and *Amphispiza bilineata.*

———
1931. Report on a collection of land birds from Sonora, Mexico. Trans. San Diego Soc. Nat. Hist. 6: 237–304.
Includes comments on Arizona specimens.

[van Rossem, A. J.]
1932. [Birds observed in southern Arizona.] [In] Minutes of Cooper Club Meetings, Condor 34: 60.

van Rossem, A. J.
1932. The avifauna of Tiburon Island, Sonora, Mexico, with descriptions of four new races. Trans. San Diego Soc. Nat. Hist. 7: 119–150.
Measurements of *Pipilo fuscus mesoleucus* of Arizona are listed.

———
1932. A southern race of the Spotted Screech Owl. Trans. San Diego Soc. Nat. Hist. 7: 183–186.
Includes range of *O. trichopsis trichopsis* in Arizona.

———
1933. The Gila Woodpecker in the Imperial Valley of California. Condor 35: 74.
Resident along the Colorado River; has recently moved westward.

———
1934. Critical notes on Middle American birds. Bull. Mus. Comp. Zool. 77: 387–490.
Includes comments on Arizona specimens.

———
1934. A record of the Cape May Warbler in Arizona. Condor 36: 165.
A specimen located in Paris, France; collected in "Arizona."

———
1936. The Bush-tit of the southern Great Basin. Auk 53: 85–86.
Remarks on Arizona specimens of *P. m. plumbeus.*

———
1936. Birds of the Charleston Mountains, Nevada. Pac. Coast Avi. 24: 1–65.
Brief comment on Arizona specimen of Broad-tailed Hummingbird.

———
1936. Notes on birds in relation to the faunal areas of south-central Arizona. Trans. San Diego Soc. Nat. Hist. 8: 121–148.
Critical notes on 42 species.

———
1939. A race of Yellow-breasted Chat from the tropical zone of southern Sonora. Wilson Bull. 51: 156.
Specimens of *I. v. auricollis* from Arizona are discussed.

———

1941. Further notes on some southwestern Yellowthroats. Condor 43: 291–292.

The range of *G. t. chryseola* is extended in Arizona to include the Altar and Santa Cruz valleys.

———

1942. Notes on some Mexican and Californian birds, with descriptions of six undescribed races. Trans. San Diego Soc. Nat. Hist. 9: 377–384.

Includes *Parabuteo unicinctus superior* and *Agelaius phoeniceus thermophilus,* both ranging into Arizona.

———

1942. The Lower California Nighthawk not a recognizable race. Condor 44: 73–74.

Measurements of *Chordeiles acutipennis texensis* from Arizona are given.

———

1942. Fuertes Red-tailed Hawk in northern Mexico and Arizona. Auk 59: 450.

In the Chiricahua Mountains.

———

1945. Preliminary studies on the Black-throated Sparrows of Baja California, Mexico. Trans. San Diego Soc. Nat. Hist. 10: 237–244.

Includes measurements of Arizona specimens of *A. b. deserticola.*

———

1945. The Golden-crowned Kinglet of southern California. Condor 47: 77–78.

Comparisons with *R. s. apache* of Arizona are made.

———

1945. The eastern distributional limits of the Anna Hummingbird in winter. Condor 47: 79–80.

Summary of migration into southern Arizona.

———

1946. An isolated colony of the Arizona Cardinal in Arizona and California. Condor 48: 247–248.

Along the Bill Williams River.

———

1947. The Dakota Song Sparrow in Arizona. Wilson Bull. 59: 38.

A specimen from Tucson.

———

1947. Comment on certain birds of Baja California, including descriptions of three new races. Proc. Biol. Soc. Wash. 60: 51–56.

Specimens of *Zenaida asiatica mearnsi* from Arizona are mentioned.

———

1947. A synopsis of the Savannah Sparrows of northwestern Mexico. Condor 49: 97–107.

Mentions Arizona specimens.

van Tyne, J.
1953. Geographical variation in the Blue-throated Hummingbird *(Lampornis clemenciae)*. Auk 70: 207–209.
Discussion of *L. c. bessophilus* with specimens from Arizona.

Visher, S. S.
1909. The capture of the Red-eyed Cowbird in Arizona. Auk 26: 307.
Tangavius aeneus involucratus near Tucson.

———
1910. A correction: A new bird for the United States. Auk 27: 210.
The Red-eyed Cowbird reported near Tucson in 1909 proves to be *T. a. aeneus*.

———
1910. Notes on the birds of Pima County, Arizona. Auk 27: 279–288.
A list of 127 species, some of them obvious misidentifications.

———
1910. A day with the birds in southern Arizona. Bird-Lore 12: 186–188.
A running account of the birds seen on June 15 from Tucson to Pima Canyon. Some of the records are questionable.

Voous, K. H., Jr.
1947. On the history of the distribution of the genus *Dendrocopus*. Limosa 20: 1–142.
Lists Arizona specimens of *D. scalaris*.

Vorhies, C. T.
1921. The Water Ouzel in Arizona. Condor 23: 131–132.
Observed in Oak Creek Canyon, along the White River, and in the Santa Catalina Mountains.

———
1930. The Rocky Mountain Pine Grosbeak in Arizona. Condor 32: 262–263.
At Jacob Lake, Kaibab Plateau.

———
1932. First record of the Pectoral Sandpiper for Arizona. Condor 34: 46–47.
On the Santa Rita Experimental Range.

———
1934. Second record of the Red-billed Tropic-bird in Arizona. Condor 36: 85–86.
One collected in Apache Pass draw between the Dos Cabezas and Chiricahua mountains.

———
1934. Records of Lesser Snow Goose and Whistling Swan in Arizona. Condor 36: 115.
Specimens of each from localities in southern Arizona.

———
1934. Arizona records of the Thick-billed Parrot. Condor 36: 180–181.
In the Chiricahua Mountains.

———

1936. Surf Scoter and Caspian Tern in Arizona. Condor 38: 248–249.
In Yavapai County.

———

1944. [Report of Black-bellied Tree Duck, Northern Flicker, and Boat-tailed Grackle in the Tucson region.] [In] Minutes of Cooper Club Meetings, Condor 46: 40.

———

1945. Another record of the Purple Gallinule in Arizona. Condor 47: 82.
Near Tucson.

———

1945. Black-bellied Tree Ducks in Arizona. Condor 47: 82.
Two collected south of Tucson.

———

1947. More records of the Wood Duck in Arizona. Condor 49: 245.

———, R. Jenks, and A. R. Phillips
1935. Bird records from the Tucson region, Arizona. Condor 37: 243–247.
Accounts of 31 species, including *Larus argentatus smithsonianus* and *Xema sabini,* new to the state list.

——— and A. R. Phillips
1937. Brown Pelicans invade Arizona. Condor 39: 175.
Observed on the Colorado River, at Mormon Lake, and near Tucson.

——— and W. P. Taylor
1935. A second occurrence of the White-fronted Goose in Arizona. Condor 37: 175.
At Parks Lake, Graham County.

Walkinshaw, L. H.
1949. The Sandhill Crane. Cranbrook Inst. Sci. Bull. 29: 1–202.
Includes summary of Arizona occurrences.

Walsh, L. L.
1933. Notes from southern Arizona. Auk 50: 124.
Trogon ambiguus, Crotophaga sulcirostris, and *Dendroica virens* are recorded.

Wauer, R. H.
1966. Bird-banding at Zion National Park. Western Bird Bander 41: 24–25.
Included because of its proximity to northwestern Arizona.

———

1968. Northern range extension of Wied's Crested Flycatcher. Condor 70: 88.
At Beaver Dam, Arizona, and Beaver Dam Wash, Utah.

———

1969. Recent bird records from the Virgin River Valley of Utah, Arizona, and Nevada. Condor 71: 331–335.
Includes many range extensions; also comments on subspecies.

——— and R. C. Russell
1967. New and additional records of birds in the Virgin River Valley. Condor 69: 420–423.
Includes several Arizona records.

Wayne, A. T.
1905. The Black-fronted Warbler *(Dendroica auduboni nigrifrons)* in southern California. Auk 22: 419.
Mentions breeding specimens from the Huachuca Mountains, Arizona.

Webster, J. D.
1959. The taxonomy of the Robin in Mexico. Wilson Bull. 71: 278–280.
Measurements of Arizona specimens are included.

——— and R. T. Orr
1958. Variation in the Great Horned Owls of middle America. Auk 75: 134–142.
Specimens of *B. v. pallescens* from Arizona examined.

Weller, M. W.
1964. Distribution and migration of the Redhead. Jour. Wildl. Mgmt. 28: 64–103.
Includes Arizona recovery records.

[Werner, S.]
1935. [Vermilion Flycatchers and Arizona Hooded Orioles observed in Phoenix, Arizona, on March 31, 1935.] [In] Minutes of Cooper Club Meetings, Condor 37: 220.

West, D. A.
1962. Hybridization in Grosbeaks *(Pheucticus)* of the Great Plains. Auk 79: 399–424.
Includes outline map of breeding range of *P. melanocephalus* that erroneously takes in most of southern Arizona.

Westcott, P. W.
1964. Invasion of Clark Nutcrackers and Pinon Jays into southeastern Arizona. Condor 66: 441.

[Western Bird-Banding Association, ed.]
1932. [L. L. Hargrave captures Western Tree Sparrow and Mexican Ground Dove in northern Arizona.] News from the Bird-Banders 7: 7.
Banding of Juncos is begun at the Museum of Northern Arizona.

———
1932. [L. L. Hargrave banded 630 birds in Arizona in 1931.] News from the Bird-Banders 7: 51.

———
1933. Additions to the western list. News from the Bird-Banders 8: 16.
One Inca Dove and six Piñon Jays banded by L. L. Hargrave.

[Western Bird-Banding Association, ed. et al]
1933. Summary of western banding, 1932. News from the Bird-Banders 8: 12–14. (Note: Titles in this series of annual reports vary. Later ones are: Annual report of birds banded in the western states for. . . .) Arizona bird species banded, and their number, are listed. 1934, 9: 13–23. 1935, 10: 1–6. 1936, 11: 13–17. 1937, 12: 14–19. 1938, 13: 11–17. 1939, 14: 14–19. 1940, 15: 12–18. 1941, 16: 12–18. 1942, 17: 15–21. 1943, 18: 5–9. 1944, 19: 9–12. 1945, 20: 7–11. 1946, 21: 8–11. 1947, 22: 21–25. 1948, 23: 14–20. 1949, 24: 20–26. 1950, 25: 16–22.

1951, 26: 16–23. 1952, 27: 14–21. 1953, 28: 14–20. 1954, 29: 14–24. 1955, 30: 13–24. 1956, 31: 16–22. 1957, 32: 13–19. 1958, 33: 9–17. 1959, 34: 15–26. 1960, 35: 13–26. (Note: Beginning with this year Arizona and New Mexico banding records are combined. As of 1961, the name of the journal is changed to Western Bird Bander.) 1961, 36: 13–24. 1962, 37: 16–31. 1963, 38: 14–25. 1964, 39: 14–24. 1965, 40: 14–28. 1966, 41: 12–22. 1967, 42: 15–26. 1968, 43: 29–36. 1969, 44: 13–22. 1970, 45: 18–26.

[Western Bird-Banding Association, ed.]

1970. Summary report of individual banders (1969) (226 responding banders). Western Bird Bander 45: 29–31.

———

1970. Correction. Western Bird Bander 45: 28.

A correction and an additional reference to the "Winter territoriality in a Ruby-crowned Kinglet" by A. M. Rea in Western Bird Bander 45: 4–7.

Wetherill, M. A.

1937. Long-tailed Chickadee in Arizona. Condor 39: 86.

First specimen for Arizona "found" in Betatakin Canyon.

——— and A. R. Phillips

1949. Bird records from the Navajo country. Condor 51: 100–102.

Accounts of 39 species.

Wetmore, A.

1908. Notes on some northern Arizona birds. Kansas Univ. Sci. Bull. 4: 377–388.

Accounts of 40 species observed in the vicinity of Williams, Arizona.

———

1921. Further notes on birds observed near Williams, Arizona. Condor 23: 60–64.

Observations on 44 species.

———

1923. Migration records from wild ducks and other birds banded in the Salt Lake Valley, Utah. U.S. Dept. Agric. Dept. Bull. 1145: 1–16.

Several Arizona recoveries are listed.

———

1933. Records from north central Arizona. Condor 35: 163 164.

Accounts of 9 species.

———

1935. The Thick-billed Parrot in southern Arizona. Condor 37: 18–21.

In the Chiricahua, Dragoon, Galiuro, and Graham Mountains; notes on habits.

Wight, H. M.

1962. Mourning Dove status report, 1962. U.S. Dept. Int. Bur. Sport Fisheries and Wildl. Spec. Sci. Rep. (Wildl.) 70: 1–33.

Includes Arizona call-counts.

——— and E. B. Baysinger

1963. Mourning Dove status report, 1963. U.S. Dept. Int. Bur. Sport Fisheries and Wildl. Spec. Sci. Rep. (Wildl.) 73: 1–34.

Wight, H. M., E. B. Baysinger, and R. E. Tomlinson
1964. Mourning Dove status report, 1964. U.S. Dept. Int. Bur. Sport Fisheries and Wildl. Spec. Sci. Rep. (Wildl.) 87: 1–38.

Willard, F. C.
1905. Notes from Cochise Co., Ariz.: Purple Gallinule. Condor 7: 112.
Ionornis martinica captured at Tombstone in June, 1904.

———
1910. Seen on a day's outing in southern Arizona. Condor 12: 110.
Ceryle americana septentrionalis collected on the San Pedro River near Fairbanks.

———
1912. A week afield in southern Arizona. Condor 14: 53–63.
Birds observed on a trip from the Huachuca Mountains to Tucson, and then to Tombstone.

———
1912. Migration of White-necked Ravens. Condor 14: 107–108.
In Cochise County in November.

———
1916. On bicycle and afoot in the Santa Catalina Mountains. Condor 18: 156–160.
A running account of birds observed in May, 1904.

———
1916. Notes on the Golden Eagle in Arizona. Condor 18: 200–201.
In the Chiricahua Mountains.

Willet, G.
1915. New winter records for Arizona. Condor 17: 102.
Erismatura jamaicensis and *Calypte anna.*

Williams, C. S.
1944. Migration of the Red-head from the Utah breeding grounds. Auk 61: 251–259.
Map shows November records from Arizona.

Wood, W.
1881. California Pygmy Owl. Ornith. and Ool. 6: 33–35; 47–48.
Observations in the White Mountains, Arizona.

Woodbury, A.M.
1939. Bird records from Utah and Arizona. Condor 41: 157–163.
Several species from Arizona are listed.

———, C. Cottam, and J. W. Sugden
1949. Annotated check-list of the birds of Utah. Univ. Utah Bull. 39 (16), Biol. Ser. 11 (2): 1–40.
Mentions records of *Buteo lineatus elegans* from northeastern Arizona.

——— and H. N. Russell, Jr.
1945. Birds of the Navajo country. Univ. Utah Bull. 35 (14), Biol. Ser. 9 (1): 1–160.
The accounts include lists of specimens, distribution, observations, and discussion. The area studied takes in parts of Utah.

——, et al
1958. Preliminary report on biological resources of the Glen Canyon Reservoir. Univ. Utah Anthro. Papers no. 31 (Glen Canyon Ser. no. 2): 1–219.
Birds: pp. 181–208. Annotated check-list by A. M. Woodbury.

Wrakestraw, G. F. and R. M. Ballou
1954. Migration patterns of Wyoming Pintails. Proc. 34th Ann. Conf. West. Assoc. State Game and Fish Comms.: 276–278.
One percent recovered in Arizona.

Wright, G. M., J. S. Dixon, and B. H. Thompson
1933. Fauna of the National Parks of the United States. U.S. Dept. Int. Nat. Park Serv. Cont. Wildl. Surv. Fauna Ser. 1: 1–157.
Brief notes on Merriam Turkey at the Grand Canyon National Park.

Zimmer, J. T.
1937. Studies of Peruvian birds, 28. Notes on the genera *Myiodynastes, Conopias, Myiozetetes,* and *Pitangus.* Amer. Mus. Novit. 963: 1–28.
Includes comments on *Myiodynastes luteiventris swarthi* from Arizona.

———
1938. Studies of Peruvian birds, 29. The genera *Myiarchus, Mitrephanes,* and *Cnemotriccus.* Amer. Mus. Novit. 994: 1–32.
Lists specimens of *Myiarchus tuberculifer olivascens* from Arizona.

———
1939. Studies of Peruvian birds, 31. Notes on the genera *Myiotriccus, Pyrrhomyias, Myiophobus, Onychorhynchus, Platyrinchus, Cnipodectes, Sayornis,* and *Nuttallornis.* Amer. Mus. Novit. 1043: 1–15.
Specimens of *Nuttallornis borealis* from Arizona; only one form is recognized.

Zimmerman, D. A.
1960. Thick-billed Kingbird nesting in New Mexico. Auk 77: 92–94.
Three pairs observed in the Arizona portion of Guadalupe Canyon, Cochise County, Arizona.

———
1965. The Gray Hawk in the southwest. Aud. Field Notes 19: 475–477.
Comments on habitat destruction.

——— and S. H. Levi
1960. Violet-crowned Hummingbird nesting in Arizona and New Mexico. Auk 77: 470–471.
In Guadalupe Canyon, Cochise County, Arizona.

Bird Counts

These inventories of species of birds, conducted in the course of a single day, under the rules of the National Audubon Society, began with the title of "Christmas Bird Censuses." Later they are reported as "Christmas Bird Counts." Although the authorship varies with the number of people participating in the counts, and may include as many as a score or more of individual observers, the name of the compiler, when available, is listed here as the author. In order to facilitate location and examination, the bird counts are listed in chronological order.

1910

Brown, H.
[Tenth Christmas bird census at] Tucson, Ariz., Dec. 25, [1909]. Bird-Lore 12: 35.

1911

Harman, W. G.
[Eleventh Christmas bird census at] Oracle, Pinal Co., Ariz., Dec. 24, [1910]. Bird-Lore 13: 41–42.

1915

Schaefer, O. F.
[Fifteenth Christmas bird census at] Flagstaff, Ariz., Dec. 20, [1914]. Bird-Lore 17: 45.

1916

Lane, J. W., Jr.
[Sixteenth Christmas bird census at] Mesa (vicinity), Maricopa Co., Ariz., Dec. 21, [1915]. Bird-Lore 18: 39–40.

1917

Schaefer, O. F.
[Seventeenth Christmas bird census at] Camp Verde, Ariz., Dec. 22, [1916]. Bird-Lore 19: 36–37.

1918

Slate, J. B.
[Eighteenth Christmas bird census at] Sacaton, Ariz., Dec. 25, [1917]. Bird-Lore 20: 48.

1919

Vorhies, C. T.
[Nineteenth Christmas bird census at] Tucson, Ariz., Dec. 26, [1918]. Bird-Lore 21: 46–47.

1923

Battles, C. D., and H. T. Kimball
[Twenty-third Christmas bird census at] Superior, Ariz., Dec. 24, [1922]. Bird-Lore 25: 43.

1928

Taylor, W. P., and C. T. Vorhies
[Twenty-eighth Christmas bird census at] Tucson, Ariz., Dec. 26, [1927]. Bird-Lore 30: 65–66.

1930

Taylor, W. P., and C. T. Vorhies
[Thirtieth Christmas bird census at] Tucson, Ariz., Dec. 24, [1929]. Bird-Lore 32: 61–62.

1931

Crockett, H. L., and Mrs. H. L. Crockett
[Thirty-first Christmas bird census at] Phoenix, Ariz., Dec. 21, [1930]. Bird-Lore 33: 74.

1932

Crockett, H. L., and Mrs. H. L. Crockett
[Thirty-second Christmas bird census at] Phoenix, Ariz., Dec. 20, [1931]. Bird-Lore 34: 75.

Phillips, A. [R.]
[Thirty-second Christmas bird census at] Baboquivari Mountains, Ariz., Dec. 20, [1931]. Bird-Lore 34: 74–75.

1933

Anderson, A. H.
[Thirty-third Christmas bird census at] Tucson, Ariz., Dec. 24, [1932]. Bird-Lore 35: 54.

1935

Monson, G.
[Thirty-fifth Christmas bird census at] Papago Indian Reservation, Ariz., Dec. 20, [1934]. Bird-Lore 37: 79–80.

1936

Monson, G., and R. J. Bryce
[Thirty-sixth Christmas bird census at] Safford, Ariz., Dec. 19, [1935]. Bird-Lore 38: 80.

Sanders, E., and H. L. and R. Crockett
[Thirty-sixth Christmas bird census at] Phoenix, Ariz., Dec. 22, [1935]. Bird-Lore 38: 80.

1937

Monson, G.
[Thirty-seventh Christmas bird census at] Tuba City, Navajo Indian Reservation, Ariz., Dec. 19, [1936]. Bird-Lore 39: 67–68.

Phillips, A. R. (compiler), et al.
[Thirty-seventh Christmas bird census at] Santa Catalina Mts., Ariz., Dec. 22, [1936]. Bird-Lore 39: 67.

1938

Phillips, A. R. (compiler), et al.
[Thirty-eighth Christmas bird census at] Santa Catalina Mts., Ariz., Dec. 26, [1937]. Bird-Lore 40: 66–67.

1939

Jackson, B.
[Thirty-ninth Christmas bird census at] Montezuma Castle National Monument, Ariz., Dec. 26, [1938]. Bird-Lore 41 (1) (supplement): 50.

Phillips, A. R. (compiler), et al.
[Thirty-ninth Christmas bird census at] Santa Catalina Mts., Ariz., Dec. 18, [1938]. Bird-Lore 41 (1) (supplement): 50–51.

1940

Phillips, A. R. (compiler), et al.
[Fortieth Christmas bird census at] Santa Catalina Mts., Ariz., Dec. 17, [1939]. Bird-Lore 42 (1) (supplement): 128.

1941

Foerster, W. X. (compiler), et al.
[Forty-first Christmas bird census at] Santa Catalina Mts., Ariz., Dec. 22, [1940]. Aud. Mag. 43 (1) (supplement): 138–139.

———
[Forty-first Christmas bird census at] Tucson, Ariz., Dec. 28, [1940]. Aud. Mag. 43 (1) (supplement): 139.

Jackson, B.
[Forty-first Christmas bird census at] Montezuma Castle National Monument, Ariz., Dec. 22, [1940]. Aud. Mag. 43 (1) (supplement): 138.

Vorhies, C. T., et al.
[Forty-first Christmas bird census at] Yuma, Ariz., Dec. 22, [1940]. Aud. Mag. 43 (1) (supplement): 139–140.

1942

Foerster, W. X. (compiler), et al.
[Forty-second Christmas bird count at] Santa Catalina Mts., Ariz., Dec. 21, [1941]. Aud. Mag. 44 (1) (supplement): 67–68.

———
[Forty-second Christmas bird count at] Tucson, Ariz., Dec. 28, [1941]. Aud. Mag. 44 (1) (supplement): 68.

Goldman, L. C., E. Sanders, and L. W. Arnold
[Forty-second Christmas bird count at] Yuma, Ariz., Dec. 22, [1941]. Aud. Mag. 44 (1) (supplement): 68.

Jackson, B.
[Forty-second Christmas bird count at] Montezuma Castle National Monument, Ariz., Dec. 27, [1941]. Aud. Mag. 44 (1) (supplement): 67.

1943

Foerster, W. X. (compiler), et al.
[Forty-third Christmas bird count at] Santa Catalina Mts., Ariz., Dec. 20, [1942]. Aud. Mag. 45 (1) sec. 2: 54–55.

———
[Forty-third Christmas bird count at] Tucson, Ariz., Dec. 27, [1942]. Aud. Mag. 45 (1) sec. 2: 55.

Monson, G.
[Forty-third Christmas bird count at] Imperial National Wildlife Refuge, Yuma, Ariz., Dec. 26, [1942]. Aud. Mag. 45 (1) sec. 2: 54.

1944

Foerster, W. X. (compiler), et al.
[Forty-fourth Christmas bird count at] Santa Catalina Mts., Ariz., Dec. 19, [1943]. Aud. Mag. 46 (1) sec. 3: 57–58.

Foerster, A. J., and W. X. Foerster
[Forty-fourth Christmas bird count at] Tucson, Ariz., Dec. 26, [1943]. Aud. Mag. 46 (1) sec. 3: 58.

Monson, G.
[Forty-fourth Christmas bird count at] Havasu Lake National Wildlife Refuge (Mohave and Yuma Cos.) Ariz., Dec. 20, [1943]. Aud. Mag. 46 (1) sec. 3: 57.

1945

Foerster, W. X. (compiler), et al.
[Forty-fifth Christmas bird count at] Santa Catalina Mts., Ariz., Dec. 17, [1944]. Aud. Mag. 47 (2) sec. 2: 57.

———
[Forty-fifth Christmas bird count at] Tucson, Ariz., Dec. 28, [1944]. Aud. Mag. 47 (2) sec. 2: 57.

Halloran, A. F., and F. B. McMurry
[Forth-fifth Christmas bird count at] Imperial National Wildlife Refuge, Ariz., Dec. 28, [1944]. Aud. Mag. 47 (2) sec. 2: 56–57.

1946

Crockett, H. L. (compiler), et al.
[Forty-sixth Christmas bird count at] Phoenix, Ariz., Dec. 29, [1945]. Aud. Mag. 48 (2) sec. 2: 85.

Foerster, W. X. (compiler), et al.
[Forty-sixth Christmas bird count at] Santa Catalina Mts., Ariz., Dec. 23, [1945]. Aud. Mag. 48 (2) sec. 2: 85.

———
[Forty-sixth Christmas bird count at] Tucson, Ariz., Dec. 30, [1945]. Aud. Mag. 48 (2) sec. 2: 85–86.

Halloran, A. F., and F. B. McMurry
[Forty-sixth Christmas bird count at] Imperial National Wildlife Refuge, Yuma, Ariz., Dec. 29, [1945]. Aud. Mag. 48 (2) sec. 2: 84–85.

1947

Halloran, A. F., and A. G. Halloran
[Forty-seventh Christmas bird count at] Imperial National Wildlife Refuge, Ariz., Dec. 28, [1946]. Aud. Field Notes 1: 108.

Monson, G.
[Forty-seventh Christmas bird count at] Havasu Lake National Wildlife Refuge (Mohave Co.), Ariz., Dec. 22, [1946]. Aud. Field Notes 1: 108.

Phillips, A. R. (compiler), et al.
[Forty-seventh Christmas bird count at] Santa Catalina Mts., Ariz., Dec. 22, [1946]. Aud. Field Notes 1 (2): 108–109.

———
[Forty-seventh Christmas bird count at] Tucson, Ariz., Dec. 29, [1946]. Aud. Field Notes 1 (2): 109.

1948

Halloran, A. F., and A. G. Halloran
[Forty-eighth Christmas bird count at] Imperial National Wildlife Refuge, Yuma, Ariz., Dec. 28, [1947]. Aud. Field Notes 2: 110.

Monson, G., and W. Pulich
[Forty-eighth Christmas bird count at] Havasu Lake National Wildlife Refuge (Mohave and Yuma Cos.), Ariz., Dec. 20, [1947]. Aud. Field Notes 2: 110.

Phillips, A. R. (compiler), et al.
[Forty-eighth Christmas bird count at] Tucson, Ariz., Dec. 20, [1947]. Aud. Field Notes 2: 111.

———
[Forty-eighth Christmas bird count at] Santa Catalina Mts., Ariz., Dec. 28, [1947]. Aud. Field Notes 2: 110–111.

1949

Halloran, A. F., and A. G. Halloran
[Forty-ninth Christmas bird count at] Imperial National Wildlife Refuge, Yuma, Ariz., Jan. 2, [1949]. Aud. Field Notes 3: 140.

Monson, G., W. Pulich, and E. S. Stergios
[Forty-ninth Christmas bird count at] Havasu Lake National Wildlife Refuge, (Mohave Co.), Ariz., Dec. 29, [1948]. Aud. Field Notes 3: 140.

Phillips, A. R. (compiler), et al.
[Forty-ninth Christmas bird count at] Santa Catalina Mts., Ariz., Jan. 2, [1949]. Aud. Field Notes 3: 140–141.

———
[Forty-ninth Christmas bird count at] Tucson, Ariz., Dec. 30, [1948]. Aud. Field Notes 3: 141.

1950

Halloran, A. F., and J. H. Gretzinger
[Fiftieth Christmas bird count at] Imperial National Wildlife Refuge, Yuma (Yuma Co.), Ariz., Dec. 31, [1949]. Aud. Field Notes 4: 164.

Monson, G., A. R. Phillips, and W. Pulich
[Fiftieth Christmas bird count at] Havasu Lake National Wildlife Refuge, (Mohave Co.), Ariz., Dec. 28, [1949]. Aud. Field Notes 4: 164.

Phillips, A. R. (compiler), et al.
[Fiftieth Christmas bird count at] Santa Catalina Mts., Ariz., Dec. 26, [1949]. Aud. Field Notes 4: 164–165.

———
[Fiftieth Christmas bird count at] Tucson, Ariz., Dec. 31, [1949]. Aud. Field Notes 4: 165–166.

1951

Halloran, A. F. (compiler), et al.
[Fifty-first Christmas bird count at] Imperial National Wildlife Refuge, Yuma, Ariz., Dec. 26, [1950]. Aud. Field Notes 5: 170.

Monson, G., and S. Stergios
[Fifty-first Christmas bird count at] Havasu Lake National Wildlife Refuge, (Mohave Co.), Ariz., Dec. 27, [1950]. Aud. Field Notes 5: 170.

Phillips, A. R. (compiler), et al.
[Fifty-first Christmas bird count at] Santa Catalina Mts., Ariz., Dec. 30, [1950]. Aud. Field Notes 5: 171.

———
[Fifty-first Christmas bird count at] Tucson, Ariz., Dec. 23, [1950]. Aud. Field Notes 5: 171–172.

———, and L. D. Yaeger
[Fifty-first Christmas bird count at] Salt River (Maricopa Co.), Ariz., Dec. 26, [1950]. Aud. Field Notes 5: 170–171.

1952

Babbitt, R. G., Mrs., et al.
[Fifty-second Christmas bird count at] Flagstaff, Ariz., Dec. 30, [1951]. Aud. Field Notes 6: 160.

Crandell, H. B., and A. F. Halloran
[Fifty-second Christmas bird count at] Imperial National Wildlife Refuge, Yuma, Ariz., Dec. 29, [1951]. Aud. Field Notes 6: 160–161.

Monson, G.
[Fifty-second Christmas bird count at] Parker, Ariz., Dec. 26, [1951]. Aud. Field Notes 6: 161.

———, E. R. Gomer, and G. R. Osbun
[Fifty-second Christmas bird count at] Havasu Lake National Wildlife Refuge, (Mohave Co.), Ariz., Dec. 23, [1951]. Aud. Field Notes 6: 160.

Peters, I. (compiler), et al.
[Fifty-second Christmas bird count at] Tucson, Ariz., Dec. 29, [1951]. Aud. Field Notes 6: 162.

Phillips, A. R. (compiler), et al.
[Fifty-second Christmas bird count at] Santa Catalina Mts., Ariz., Dec. 23, [1951]. Aud. Field Notes 6: 161–162.

1953

McLernon, H. (compiler), and U. Miller
[Fifty-third Christmas bird count at] Prescott, Ariz., Dec. 28, [1952]. Aud. Field Notes 7: 175.

Monson, G.
[Fifty-third Christmas bird count at] Havasu Lake National Wildlife Refuge, Topock, Ariz., Dec. 28, [1952]. Aud. Field Notes 7: 174–175.

———
[Fifty-third Christmas bird count at] Parker, Ariz., Dec. 24, [1952]. Aud. Field Notes 7: 175.

Peters, I. (compiler), et al.
[Fifty-third Christmas bird count at] Tucson, Ariz., Dec. 27, [1952]. Aud. Field Notes 7: 175–176.

Pugh, R. W., Mr. and Mrs. (compilers), et al.
[Fifty-third Christmas bird count at] Flagstaff, Ariz., Dec. 28, [1952]. Aud. Field Notes 7: 174.

1954

McLernon, H.
[Fifty-fourth Christmas bird count at] Prescott, Ariz., Dec. 31, [1953]. Aud. Field Notes 8: 208.

Monson, G. (compiler), et al.
[Fifty-fourth Christmas bird count at] Havasu Lake National Wildlife Refuge, Topock, Ariz., Dec. 29, [1953]. Aud. Field Notes 8: 207.

———, and C. R. Darling
[Fifty-fourth Christmas bird count at] Parker, Ariz., Dec. 31, [1953]. Aud. Field Notes 8: 207–208.

Peters, I. (compiler), et al.
[Fifty-fourth Christmas bird count at] Tucson, Ariz., Jan. 2, [1954]. Aud. Field Notes 8: 209.

Phillips, A. R. (compiler), et al.
[Fifty-fourth Christmas bird count at] Santa Catalina Mts., Ariz., Dec. 26, [1953]. Aud. Field Notes 8: 208–209.

Pugh, R. W. Mrs. (compiler), et al.
[Fifty-fourth Christmas bird count at] Flagstaff, Ariz., Jan. 3, [1954]. Aud. Field Notes 8: 207.

Sutton, E. A. (compiler), et al.
[Fifty-fourth Christmas bird count at] Verde Valley, Ariz., Dec. 31, [1953]. Aud. Field Notes 8: 209.

1955

Johnson, H. J., and G. Monson
[Fifty-fifth Christmas bird count at] Imperial National Wildlife Refuge, Ariz.-Calif., Dec. 29, [1954]. Aud. Field Notes 9: 214.

MacArthur, R. H., and E. W. MacArthur
[Fifty-fifth Christmas bird count at] Hereford, Ariz., Dec. 29, [1954]. Aud. Field Notes 9: 214.

Margolin, A. S. (compiler), et al.
[Fifty-fifth Christmas bird count at] Phoenix, Ariz., Dec. 27, [1954]. Aud. Field Notes 9: 214–215.

———
[Fifty-fifth Christmas bird count at] Salt River, Ariz., Dec. 30, [1954]. Aud. Field Notes 9: 215.

McLernon, H. (compiler), L. Ebe, and U. Miller
[Fifty-fifth Christmas bird count at] Prescott, Ariz., Jan. 2, [1955]. Aud. Field Notes 9: 215.

Monson, G.
[Fifty-fifth Christmas bird count at] Yuma, Ariz., Jan. 1, [1955]. Aud. Field Notes 9: 216.

Peters, I. (compiler), et al.
[Fifty-fifth Christmas bird count at] Santa Catalina Mts., Dec. 31, [1954]. Aud. Field Notes 9: 215–216.

1956

Margolin, A. S. (compiler), et al.
[Fifty-sixth Christmas bird count at] Phoenix, Ariz., Dec. 26, [1955]. Aud. Field Notes 10: 209–210.

McLernon, H.
[Fifty-sixth Christmas bird count at] Prescott, Ariz., Jan. 1, [1956]. Aud. Field Notes 10: 210.

Monson, G.
[Fifty-sixth Christmas bird count at] Sierra Pinta, Cabeza Prieta Game Range, Ariz., Dec. 30, [1955]. Aud. Field Notes 10: 210–211.

Monson, G. (compiler), and H. D. Irby
[Fifty-sixth Christmas bird count at] Imperial National Wildlife Refuge, Ariz.-Calif., Dec. 27, [1955]. Aud. Field Notes 10: 209.

Phillips, A. R. (compiler), et al.
[Fifty-sixth Christmas bird count at] Santa Catalina Mts., Ariz., Dec. 31, [1955]. Aud. Field Notes 10: 210.

1957

Margolin, A. S. (compiler), et al.
[Fifty-seventh Christmas bird count at] Phoenix, Ariz., Dec. 30, [1956]. Aud. Field Notes 11: 216.

Monson, G.
[Fifty-seventh Christmas bird count at] Imperial National Wildlife Refuge, Ariz.-Calif., Dec. 29, [1956]. Aud. Field Notes 11: 215–216.

———
[Fifty-seventh Christmas bird count at] Sierra Pinta, Cabeza Prieta Game Range, Ariz., Dec. 27, [1956]. Aud. Field Notes 11: 216–217.

Woodard, R. E. (compiler), et al.
[Fifty-seventh Christmas bird count at] Santa Catalina Mts., Ariz., Dec. 26, [1956]. Aud. Field Notes 11: 216.

———
[Fifty-seventh Christmas bird count at] Tucson, Ariz., Dec. 29, [1956]. Aud. Field Notes 11: 217.

1958

Margolin, A. S. (compiler), et al.
[Fifty-eighth Christmas bird count at] Phoenix, Ariz., Dec. 29, [1957]. Aud. Field Notes 12: 226.

McLernon, H. (compiler), and U. Miller
[Fifty-eighth Christmas bird count at] Prescott, Ariz., Dec. 22, [1957]. Aud. Field Notes 12: 226–227.

Monson, G.
[Fifty-eighth Christmas bird count at] Sierra Pinta, Cabeza Prieta Game Range, Ariz., Dec. 27, [1957]. Aud. Field Notes 12: 227.

Monson, G. (compiler), J. T. Bialac, and R. R. Johnson
[Fifty-eighth Christmas bird count at] Imperial National Wildlife Refuge, Ariz.-Calif., Dec. 30, [1957]. Aud. Field Notes 12: 226.

Stimson, L. A. (compiler), et al.
[Fifty-eighth Christmas bird count at] Santa Catalina Mts., Ariz., Dec. 21, [1957]. Aud. Field Notes 12: 227.

———
[Fifty-eighth Christmas bird count at] Tucson, Ariz., Dec. 28, [1957]. Aud. Field Notes 12: 227–228.

1959

Margolin, A. S. (compiler), et al.
[Fifty-ninth Christmas bird count at] Phoenix, Ariz., Dec. 21, [1958]. Aud. Field Notes 13: 238–239.

McLernon, H. (compiler), and O. A. Humphrey
[Fifty-ninth Christmas bird count at] Prescott, Ariz., Dec. 30, [1958]. Aud. Field Notes 13: 239.

Monson, G.
[Fifty-ninth Christmas bird count at] Imperial National Wildlife Refuge, Ariz.-Calif., Dec. 20, [1958]. Aud. Field Notes 13: 238.

Monson, G. (compiler), and Mr. and Mrs. L. Hilton
[Fifty-ninth Christmas bird count at] Sierra Pinta, Cabeza Prieta Game Range, Ariz., Dec. 30, [1958]. Aud. Field Notes 13: 239–240.

Peters, I. (compiler), et al.
[Fifty-ninth Christmas bird count at] Santa Catalina Mts., Ariz., Dec. 20, [1958]. Aud. Field Notes 13: 239.

———
[Fifty-ninth Christmas bird count at] Tucson, Ariz., Dec. 27, [1958]. Aud. Field Notes 13: 240.

1960

Demaree, S. R. (compiler), et al.
[Sixtieth Christmas bird count at] Phoenix, Ariz., Jan. 3, [1960]. Aud. Field Notes 14: 254.

Harrison, B. [=W.] (compiler), and D. Harrison
[Sixtieth Christmas bird count at] Nogales, Ariz., Dec. 23, [1959]. Aud. Field Notes 14: 254.

McLernon, H. (compiler), and D. and S. Feese
[Sixtieth Christmas bird count at] Prescott, Ariz., Jan. 3, [1960]. Aud. Field Notes 14: 254–255.

Monson, G.
[Sixtieth Christmas bird count at] Sierra Pinta, Cabeza Prieta Game Range, Ariz., Dec. 30, [1959]. Aud. Field Notes 14: 256.

Monson, G. (compiler), and B. J. Van Tries
[Sixtieth Christmas bird count at] Imperial National Wildlife Refuge, Ariz.-Calif., Dec. 23, [1959]. Aud. Field Notes 14: 253–254.

Peters, I., and F. Thornburg (compilers), et al.
[Sixtieth Christmas bird count at] Santa Cruz Valley, Tucson, Ariz., Jan. 2, [1960]. Aud. Field Notes 14: 255–256.

Thornburg, F. (compiler), et al.
[Sixtieth Christmas bird count at] Santa Catalina Mts., Ariz., Dec. 26, [1959]. Aud. Field Notes 14: 255.

1961

Demaree, S. R. (compiler), et al.
[Sixty-first Christmas bird count at] Phoenix, Ariz., Dec. 26, [1960]. Aud. Field Notes 15: 270.

Harrison, B. [= W.] (compiler), et al.
[Sixty-first Christmas bird count at] Atascosa Peak, Ariz., Dec. 23, [1960]. Aud. Field Notes 15: 269.

Harrison, W. I. (compiler), et al.
[Sixty-first Christmas bird count at] Nogales, Ariz., Dec. 30, [1960]. Aud. Field Notes 15: 270.

Monson, G. (compiler), et al.
[Sixty-first Christmas bird count at] Imperial National Wildlife Refuge, Ariz.-Calif., Dec. 23, [1960]. Aud. Field Notes 15: 269–270.

Monson, G. (compiler), and F. Monson
[Sixty-first Christmas bird count at] Sierra Pinta, Cabeza Prieta Game Refuge, Ariz., Dec. 30, [1960]. Aud. Field Notes 15: 271–272.

Peters, I., and F. Thornburg (compilers), et al.
[Sixty-first Christmas bird count at] Santa Catalina Mts., Ariz., Dec. 28, [1960]. Aud. Field Notes 15: 271.

———
[Sixty-first Christmas bird count at] Santa Cruz Valley, Ariz., Dec. 21, [1960]. Aud. Field Notes 15: 271.

1962

Demaree, S. R. (compiler), et al.
[Sixty-second Christmas bird count at] Phoenix, Ariz., Dec. 30, [1961]. Aud. Field Notes 16: 269–270.

Harrison, B. [= W.] (compiler), et al.
[Sixty-second Christmas bird count at] Atascosa Highlands, Ariz., Dec. 30, [1961]. Aud. Field Notes 16: 268.

Harrison, B. [= W.] (compiler), et al.
[Sixty-second Christmas bird count at] Nogales, Ariz., Jan. 1, [1962]. Aud. Field Notes 16: 269.

Monson, G. (compiler), et al.
[Sixty-second Christmas bird count at] Imperial National Wildlife Refuge, Ariz.-Calif., Dec. 23, [1961]. Aud. Field Notes 16: 268–269.

———, and F. Monson
[Sixty-second Christmas bird count at] Sierra Pinta, Cabeza Prieta Game Range, Ariz., Dec. 27, [1961]. Aud. Field Notes 16: 270.

Tainter, F. (compiler), et al.
[Sixty-second Christmas bird count at] Santa Catalina Mts., Ariz., Dec. 21, [1961]. Aud. Field Notes 16: 270.

———
[Sixty-second Christmas bird count at] Tucson, Ariz., Dec. 23, [1961]. Aud. Field Notes 16: 270–271.

1963

Demaree, S. R. (compiler), et al.
[Sixty-third Christmas bird count at] Phoenix, Ariz., Dec. 29, [1962]. Aud. Field Notes 17: 265–266.

Douglass, J. (compiler), and M. Douglass
[Sixty-third Christmas bird count at] Petrified Forest National Park, Ariz., Dec. 21, [1962]. Aud. Field Notes 17: 265.

Harrison, W. (compiler), et al.
[Sixty-third Christmas bird count at] Atascosa Highlands, Ariz., Dec. 20, [1962]. Aud. Field Notes 17: 264.

———
[Sixty-third Christmas bird count at] Nogales, Ariz., Dec. 28, [1962]. Aud. Field Notes 17: 264–265.

———, and F. Thornburg (compilers), et al.
[Sixty-third Christmas bird count at] Patagonia, Ariz., Dec. 26, [1962]. Aud. Field Notes 17: 265.

Tainter, F. (compiler), et al.
[Sixty-third Christmas bird count at] Tucson, Ariz., Dec. 29, [1962]. Aud. Field Notes 17: 266–267.

Thornburg, F. (compiler), et al.
[Sixty-third Christmas bird count at] Santa Catalina Mts., Ariz., Dec. 22, [1962]. Aud. Field Notes 17: 266.

1964

Demaree, S. R. (compiler), et al.
[Sixty-fourth Christmas bird count at] Phoenix, Ariz., Dec. 28, [1963]. Aud. Field Notes 18: 295.

Douglass, J.
[Sixty-fourth Christmas bird count at] Petrified Forest National Park, Ariz., Jan. 1, [1964]. Aud. Field Notes 18: 295.

Harrison, B. [= W.] (compiler), D. Carter, and W. Winslow
[Sixty-fourth Christmas bird count at] Atascosa Highlands, Ariz., Dec. 28, [1963]. Aud. Field Notes 18: 293–294.

Harrison, B. [= W.] (compiler), et al.
[Sixty-fourth Christmas bird count at] Nogales, Ariz., Dec. 30, [1963]. Aud. Field Notes 18: 294.

———
[Sixty-fourth Christmas bird count at] Patagonia, Ariz., Dec. 26, [1963]. Aud. Field Notes 18: 294–295.

Lund, B.
[Sixty-fourth Christmas bird count at] Tucson Mountains, Ariz., Dec. 20, [1963]. Aud. Field Notes 18: 297.

Lund, B. (compiler), and W. Dengler
[Sixty-fourth Christmas bird count at] Rincon Mountains, Ariz., Dec. 27, [1963]. Aud. Field Notes 18: 296.

Tainter, F. (compiler), et al.
[Sixty-fourth Christmas bird count at] Santa Catalina Mts., Ariz., Dec. 21, [1963]. Aud. Field Notes 18: 296.

———
[Sixty-fourth Christmas bird count at] Tucson, Ariz., Dec. 28, [1963]. Aud. Field Notes 18: 297.

Wauer, R.
[Sixty-fourth Christmas bird count at] Pipe Springs National Monument, Ariz., Dec. 23, [1963]. Aud. Field Notes 18: 295–296.

1965

Demaree, S. R. (compiler), et al.
[Sixty-fifth Christmas bird count at] Phoenix, Ariz., Dec. 26, [1964]. Aud. Field Notes 19: 313.

Fultz, R. (compiler), et al.
[Sixty-fifth Christmas bird count at] Chiricahua Mts., Ariz., Dec. 22, [1964]. Aud. Field Notes 19: 311–312.

Harrison, B. [= W.] (compiler), et al.
[Sixty-fifth Christmas bird count at] Atascosa Highlands, Ariz., Dec. 28, [1964]. Aud. Field Notes 19: 311.

———
[Sixty-fifth Christmas bird count at] Nogales, Ariz., Dec. 30, [1964]. Aud. Field Notes 19: 312.

———
[Sixty-fifth Christmas bird count at] Patagonia, Ariz., Jan. 2, [1965]. Aud. Field Notes 19: 312–313.

Lund, B. (compiler), et al.
[Sixty-fifth Christmas bird count at] Rincon Mountains, Ariz., Dec. 22, [1964]. Aud. Field Notes 19: 313–314.

———
[Sixty-fifth Christmas bird count at] Tucson Mountains, Ariz., Dec. 31, [1964]. Aud. Field Notes 19: 315.

Tainter, F. (compiler), et al.
[Sixty-fifth Christmas bird count at] Santa Catalina Mts., Ariz., Dec. 23, [1964]. Aud. Field Notes 19: 314.

———
[Sixty-fifth Christmas bird count at] Tucson, Ariz., Jan. 2, [1965]. Aud. Field Notes 19: 314–315.

Wauer, R. (compiler), and A. Hagood
[Sixty-fifth Christmas bird count at] Pipe Spring National Monument, Ariz., Dec. 22, [1964]. Aud. Field Notes 19: 313.

1966

Demaree, S. R. (compiler), et al.
[Sixty-sixth Christmas bird count at] Phoenix, Ariz., Dec. 26, [1965]. Aud. Field Notes 20: 353–354.

Harrison, B. [= W.] (compiler), et al.
[Sixty-sixth Christmas bird count at] Nogales, Ariz., Dec. 21, [1965]. Aud. Field Notes 20: 352–353.

———
[Sixty-sixth Christmas bird count at] Patagonia, Ariz., Dec. 23, [1965]. Aud. Field Notes 20: 353.

———, and B. Lund
[Sixty-sixth Christmas bird count at] Atascosa Highlands, Ariz., Dec. 22, [1965]. Aud. Field Notes 20: 352.

Hoy, W. (compiler), et al.
[Sixty-sixth Christmas bird count at] Organ Pipe Cactus National Monument, Ariz., Dec. 27, [1965]. Aud. Field Notes 20: 353.

Lund, B. (compiler), et al.
[Sixty-sixth Christmas bird count at] Rincon Mountains, Ariz., Dec. 28, [1965]. Aud. Field Notes 20: 354.

———
[Sixty-sixth Christmas bird count at] Tucson Mountains, Ariz., Dec. 29, [1965]. Aud. Field Notes 20: 355.

Schaefer, J. (compiler), et al.
[Sixty-sixth Christmas bird count at] Tucson Valley, Ariz., Dec. 27, [1965]. Aud. Field Notes 20: 355.

Sheppard, J. (compiler), and L. Sheppard
[Sixty-sixth Christmas bird count at] Lower Colorado River, Calif.-Ariz., Dec. 29, [1965]. Aud. Field Notes 20: 368.

Thornburg, F. (compiler), et al.
[Sixty-sixth Christmas bird count at] Santa Catalina Mts., Ariz., Dec. 30, [1965]. Aud. Field Notes 20: 354–355.

Wauer, R. (compiler), and A. Hagood
[Sixty-sixth Christmas bird count] at Pipe Springs National Monument, Ariz., Dec. 26, [1965]. Aud. Field Notes 20: 354.

1967

Caldwell, M. (compiler), et al.
[Sixty-seventh Christmas bird count at] Rincon Mountains, Ariz., Dec. 31, [1966]. Aud. Field Notes 21: 350–351.

Caldwell, M. (compiler), and B. Lund
[Sixty-seventh Christmas bird count at] Tucson Mountains, Ariz., Dec. 22, [1966]. Aud. Field Notes 21: 351–352.

Cunningham, R. (compiler), et al.
[Sixty-seventh Christmas bird count at] Organ Pipe Cactus National Monument, Ariz., Dec. 27, [1966]. Aud. Field Notes 21: 349.

Demaree, S. R. (compiler), et al.
[Sixty-seventh Christmas bird count at] Phoenix, Ariz., Dec. 26, [1966]. Aud. Field Notes 21: 350.

Harrison, B. [= W.] (compiler), et al.
[Sixty-seventh Christmas bird count at] Nogales, Ariz., Dec. 21, [1966]. Aud. Field Notes 21: 348–349.

———
[Sixty-seventh Christmas bird count at] Atascosa Highlands, Ariz., Dec. 22, [1966]. Aud. Field Notes 21: 348.

———
[Sixty-seventh Christmas bird count at] Patagonia, Ariz., Dec. 23, [1966]. Aud. Field Notes 21: 349–350.

Hoy, W. (compiler), et al.
[Sixty-seventh Christmas bird count at] Ajo Mountains, Ariz., Dec. 21, [1966]. Aud. Field Notes 21: 347–348.

Lund, B. (compiler), and H. Bozarth
[Sixty-seventh Christmas bird count at] Pipe Spring National Monument, Ariz., Dec. 28, [1966]. Aud. Field Notes 21: 350.

Schaefer, J. (compiler), et al.
[Sixty-seventh Christmas bird count at] Tucson Valley, Ariz., Dec. 29, [1966]. Aud. Field Notes 21: 352.

Thornburg, F. (compiler), et al.
[Sixty-seventh Christmas bird count at] Santa Catalina Mts., Ariz., Dec. 26, [1966]. Aud. Field Notes 21: 351.

1968

Balda, R. (compiler), et al.
[Sixty-eighth Christmas bird count at] Flagstaff, Ariz., Dec. 30, [1967]. Aud. Field Notes 22: 362.

Cunningham, R. (compiler), et al.
[Sixty-eighth Christmas bird count at] Organ Pipe Cactus National Monument, Ariz., Dec. 29, [1967]. Aud. Field Notes 22: 363.

Demaree, S. R. (compiler), et al.
[Sixty-eighth Christmas bird count at] Phoenix, Ariz., Dec. 30, [1967]. Aud. Field Notes 22: 364.

Harrison, B. [=W.] (compiler), and E. Willis
[Sixty-eighth Christmas bird count at] Atascosa Highlands, Ariz., Dec. 20, [1967]. Aud. Field Notes 22: 361–362.

Harrison, B. [=W.] (compiler), et al.
[Sixty-eighth Christmas bird count at] Nogales, Ariz., Dec. 21, [1967]. Aud. Field Notes 22: 362–363.

———
[Sixty-eighth Christmas bird count at] Patagonia, Ariz., Dec. 22, [1967]. Aud. Field Notes 22: 363–364.

Hill, W. (compiler), et al.
[Sixty-eighth Christmas bird count at] Grand Canyon, Ariz., Jan. 1, [1968]. Aud. Field Notes 22: 362.

Lund, B. (compiler), and J. Schaack
[Sixty-eighth Christmas bird count at] Pipe Springs National Monument, Ariz., Dec. 24, [1967]. Aud. Field Notes 22: 364–365.

Manire, E. (compiler), et al.
[Sixty-eighth Christmas bird count at] Rincon Mountains, Ariz., Dec. 24, [1967]. Aud. Field Notes 22: 365.

Schaefer, J. (compiler), et al.
[Sixty-eighth Christmas bird count at] Santa Catalina Mts., Ariz., Dec. 30, [1967]. Aud. Field Notes 22: 365.

———
[Sixty-eighth Christmas bird count at] Tucson Valley, Ariz., Dec. 27, [1967]. Aud. Field Notes 22: 365–366.

Taylor, J. (compiler), et al.
[Sixty-eighth Christmas bird count at] Ajo Mountains, Ariz., Dec. 26, [1967]. Aud. Field Notes 22: 362.

1969

Balda, R. (compiler), et al.
[Sixty-ninth Christmas bird count at] Flagstaff, Ariz., Dec. 28, [1968]. Aud. Field Notes 23: 389–390.

Caldwell, M. (compiler), et al.
[Sixty-ninth Christmas bird count at] Tucson Mountains, Ariz., Dec. 29, [1968]. Aud. Field Notes 23: 395.

Collister, A. (compiler), and R. Krieger
[Sixty-ninth Christmas bird count at] Martinez Lake, Ariz., Dec. 31, [1968]. Aud. Field Notes 23: 391.

Cunningham, R. (compiler), et al.
[Sixty-ninth Christmas bird count at] Lukeville, Ariz., Dec. 23, [1968]. Aud. Field Notes 23: 390–391.

———
[Sixty-ninth Christmas bird count at] Organ Pipe Cactus National Monument, Ariz., Dec. 30, [1968]. Aud. Field Notes 23: 392.

Demaree, S. R. (compiler), et al.
[Sixty-ninth Christmas bird count at] Phoenix, Ariz., Dec. 29, [1968]. Aud. Field Notes 23: 392–393.

Gates, J. (compiler), et al.
[Sixty-ninth Christmas bird count at] Santa Catalina Mts., Ariz., Dec. 28, [1968]. Aud. Field Notes 23: 394–395.

———
[Sixty-ninth Christmas bird count at] Tucson Valley, Ariz., Dec. 26, [1968]. Aud. Field Notes 23: 395.

Harrison, B. [= W.] (compiler), et al.
[Sixty-ninth Christmas bird count at] Atascosa Highlands, Ariz., Dec. 20, [1968]. Aud. Field Notes 23: 389.

———
[Sixty-ninth Christmas bird count at] Nogales, Ariz., Dec. 21, [1968]. Aud. Field Notes 23: 391–392.

———
[Sixty-ninth Christmas bird count at] Patagonia, Ariz., Dec. 22, [1968]. Aud. Field Notes 23: 392.

Hill, W. (compiler), et al.
[Sixty-ninth Christmas bird count at] Grand Canyon, Ariz., Jan. 1, [1969]. Aud. Field Notes 23: 390.

Manire, E. (compiler), et al.
[Sixty-ninth Christmas bird count at] Rincon Mountains, Ariz., Dec. 22, [1968]. Aud. Field Notes 23: 394.

Sillick, F. (compiler), et al.
[Sixty-ninth Christmas bird count at] Ramsey Canyon, Ariz., Jan. 1, [1969]. Aud. Field Notes 23: 393.

3. Paleontology and Archaeology

Reports of fossil bird bones in Arizona are not numerous, and papers on remains of birds found in prehistoric Indian dwellings are widely scattered in many diverse publications.

Adams, W. Y., A. J. Lindsay, Jr., and C. G. Turner II

1961. Survey and excavations in Lower Glen Canyon 1952–1958. Mus. N. Ariz. Bull. 36: 1–62.

Bones of Turkey, Owl, Loon, and Gambel's Quail found.

Brodkorb, P.

1963. Catalogue of fossil birds. Bull. Florida State Mus. Biol. Sci. 7: 179–293.

Includes several Arizona specimens.

———

1964. Catalogue of fossil birds. Part 2 (Anseriformes through Galliformes). Bull. Florida State Mus. Biol. Sci. 8: 195–335.

Lists several Arizona species.

———

1964. Notes on fossil Turkeys. Quart. J. Florida Acad. Sci. 27: 223–229.

Page 225: An Arizona specimen from vicinity of Benson is "apparently referable" to *Agriocharis progenes*.

———

1967. Catalogue of fossil birds: 3 (Ralliformes, Ichthyornithiformes, Charadriiformes). Bull. Florida State Mus. Biol. Sci. 11: 99–220.

Includes *Rallus phillipsi* from Arizona.

Caywood, L. R. and E. H. Spicer

1935. Tuzigoot. The excavation and repair of a ruin on the Verde River near Clarkdale, Arizona. Field Div. of Education, Nat. Park Serv. Berkeley, Calif. 119 pp.

Bones of 12 species of birds found.

Cracraft, J.

1968. The Whooping Crane from the lower Pleistocene of Arizona. Wilson Bull. 80: 490.

In San Simon Valley.

de Laguna, F.
1942. The Bryn Mawr dig at Cinder Park, Arizona. Plateau 14: 53–56.
Turkey bones were found.

deSaussure, R.
1956. Remains of the California Condor in Arizona caves. Plateau 29: 44–45.
Fossil remains.

Di Peso, C. C.
1951. The Babocomari Village site on the Babocomari River, southeastern Arizona. Amerind Found. 5: 1–248.
Bird bones found: *Aquila, Buteo, Bubo, Anas*.

———
1958. The Reeve Ruin of southeastern Arizona. Amerind Found. 8: 1–186.
Mentions finding bones of several species of birds.

Feduccia, J. A.
1968. The Pliocene Rails of North America. Auk 85: 441–453.
Includes remarks on *Rallus phillipsi* from Arizona.

Fewkes, J. W.
1904. Two summers' work in Pueblo Ruins. In 22nd annual report of the Bur. Amer. Ethno. 1900–1901, part I, pp. 1–195.
Bird bones of 9 species are listed (p. 110).

Gazin, C. L.
1942. The late Cenozoic vertebrate faunas from the San Pedro Valley, Ariz. Proc. U.S. Nat. Mus. 92: 475–518.
Contains a list of the birds.

Gidley, J. W.
1922. Preliminary report on fossil vertebrates of the San Pedro Valley, Arizona, with descriptions of new species of Rodentia and Lagomorpha. U.S. Geol. Surv. Prof. Paper 131-E, pp. 119–131.
Lists 11 species of birds.

———
1922. Field explorations in the San Pedro Valley and Sulphur Springs Valley of southern Arizona. Smiths. Misc. Coll. 72 (15) Publ. 2669: 25–30.
A report of bird fossils found.

Gilmore, C. W.
1941. A history of the Division of Vertebrate Paleontology in the United States National Museum. Proc. U.S. Nat. Mus. 90 (3109): 305–377.
Arizona fossil collections are listed.

Gladwin, H. S., E. W. Haury, E. B. Sayles, and N. Gladwin
1937. Excavations at Snaketown. Material culture. Medallion Papers 25: 1–305.
Bones of Raven, Golden Eagle, Turkey, Sage Hen, Blue Goose, and Owl sp. were found.

Guernsey, S. J.

1931. Explorations in northeastern Arizona. Report on the archeological fieldwork of 1920–1923. Papers Peabody Mus. Amer. Archeol. and Ethnol. Harvard Univ. 12 (1): 1–123.

Evidence of Turkeys was found.

Hargrave, L. L.

1939. Bird bones from abandoned Indian dwellings in Arizona and Utah. Condor 41: 206–210.

Thirty species are identified.

———

1965. Archaeological bird bones from Chapin Mesa, Mesa Verde National Park. Amer. Antiquity 31: 156–160.

———

1965. Turkey bones from Wetherill Mesa. Amer. Antiquity 31: 161–166.

Includes comments on Arizona Turkey distribution.

———

1965. Identification of feather fragments by microstudies. Amer. Antiquity 31: 202–205.

Chiefly techniques used in Glen Canyon studies.

———

1969. Bird remains from vicinity of Navajo National Monument, Arizona, pp. 67–68. [In] K. M. Anderson, Archeology on the Shonto Plateau, northeast Arizona. S. W. Mon. Assoc. Tech. Ser. 7: 1–68.

Bones of Turkey and Quail sp. ? found.

———

1970. Mexican Macaws: Comparative Osteology. Univ. Ariz. Press, Tucson. ix + 67 pp.

Macaw remains recovered from archaeological sites in Arizona and New Mexico.

Haury, E. W.

1932. Roosevelt: 9:6. A Hohokam site of the colonial period. Medallion Papers 11: 1–134.

Mentions a shell ornament carved in the shape of a Pelican. Pelicans were often painted on pottery.

———

1934. The Canyon Creek ruin and the cliff dwellings of the Sierra Ancha. Medallion Papers 14: 1–173.

Mentions wild Turkey bones found in refuse.

———

1940. Excavations in the Forestdale Valley, east-central Arizona, with an appendix: The skeletal remains of the Bear Ruin by N. E. Gabel. Univ. Ariz. Soc. Sci. Bull. 12: 1–147.

Bones of Turkey and Horned Owl found.

———

1945. Painted Cave. Northeastern Arizona. Amerind Found. 3: 1–87.

Mentions finding a Turkey pen with three eggs.

———

1950. The stratigraphy and archaeology of Ventana Cave, Arizona. Univ. Ariz. Press, Tucson. 599 pp.

Bones of 13 species of birds found.

Hay, O. P.

1927. The Pleistocene of the western region of North America and its vertebrated animals. Carnegie Inst. Wash. Publ. 322 B: 1–346.

Brief mention of Arizona birds on pp. 136–137: a summary of Gidley's 1922 paper.

Hough, W.

1903. Archeological field work in northeastern Arizona. The Museum-Gates expedition of 1901. U.S. Nat. Mus. Report, 1901: 279–358.

Lists bird remains; the Turkey was domesticated.

———

1914. Culture of the ancient pueblos of the upper Gila River region, New Mexico and Arizona. Second Museum-Gates expedition. U.S. Nat. Mus. Bull. 87: 1–139.

Includes bird remains in caves, and use of birds and bird symbols in religious observances.

———

1930. Exploration of ruins in the White Mountain Apache Indian Reservation, Arizona. Proc. U.S. Nat. Mus. 78: art. 13, no. 2856: 1–21.

Includes a list of the bird bones found in the debris of the ruins.

Howard, H.

1931. A new species of Road-runner from Quaternary cave deposits in New Mexico. Condor 33: 206–209.

Specimens of *G. californianus* from Arizona examined.

James, G. W.

1905. The Indians of the Painted Desert region. Little, Brown and Co., Boston. 268 pp.

Feathers of several species of birds were used in religious ceremonies.

Judd, N. M.

1930. The excavation and repair of Betatakin. Proc. U.S. Nat. Mus. 77 (5): 1–77.

Lists two Turkey bones, one of which was fashioned into an awl.

Kidder, A. V. and S. J. Guernsey

1919. Archeological explorations in northeastern Arizona. Smiths. Inst. Bur. Amer. Ethn. Bull. 65: 1–228.

Turkey bones found in the Kayenta area.

King, D. S.

1949. Nalakihu. Excavations at Pueblo III site on Wupatki National Monument. Mus. N. Ariz. Bull. 23: 1–183.

Lists bones of Burrowing, Long-eared, Horned, and Short-eared Owls; American Raven; and Military Macaw.

Lockett, H. G.

1933. The unwritten literature of the Hopi. Univ. Ariz. Bull. 4 (4): 1–101.

Brief notes on use of bird feathers.

Lockett, H. C. and L. L. Hargrave
1953. Woodchuck Cave. Mus. N. Ariz. Bull. 26: 1–33.
Bones of Turkey, Horned Owl, and unidentified *Passeriformes* were found.

Long, P. V., Jr.
1966. Archeological excavations in Lower Glen Canyon, Utah, 1959–1960. Mus. N. Ariz. Bull. 42: 1–80.
In Utah they found bones of Willet, Horned Owl, Common Merganser, Eared Grebe, and Belted Kingfisher. Included here because of its proximity to Arizona.

McGregor, J. C.
1937. Winona Village. A XIIth century settlement with a ball court near Flagstaff, Arizona. Mus. N. Ariz. Bull. 12: 1–53.
Bones of Mourning Dove were found.

———
1941. Winona and Ridge Ruin. Part I. Architecture and material culture. Mus. N. Ariz. Bull. 18: 1–313.
Bones of 24 species of birds listed.

Miller, A. H.
1932. Bird remains from Indian dwellings in Arizona. Condor 34: 138–139.
Eleven species identified.

———
1940. An early record of the Dickcissel in Arizona. Condor 42: 125.
A maxilla found among other bird bones at an Indian dwelling near Winona Village, Coconino County.

Miller, L.
1960. Condor remains from Rampart Cave, Arizona. Condor 62: 70.

———
1960. Correction. Condor 62: 298.
His report (Condor 62, 1960: 70) is the first of fossil remains of the Condor, but not of Recent remains.

Murray, B. G., Jr.
1967. Grebes from the late Pliocene of North America. Condor 69: 277–288.
The unnamed Grebe specimen from Arizona (Wetmore, 1924) may be *Podiceps discors.*

Parmalee, P. W.
1969. California Condor and other birds from Stanton Cave, Arizona. Jour. Ariz. Acad. Sci. 5: 204–206.
At the Grand Canyon: 13 species of birds from the bone samples.

Reed, E. K.
1951. Turkeys in SW archeology. El Palacio 58: 195–205.
A good summary.

Rixey, R. and C. B. Voll
1962. Archaeological materials from Walnut Canyon cliff dwellings. Plateau 34: 85–96.
Bones of Turkey found.

Roberts, F. H. H., Jr.
1940. Archeological remains in the Whitewater district eastern Arizona. Part II. Artifacts and Burials. Smiths. Inst. Bur. Amer. Eth. Bull. 126: 1–170.
Bones of Turkey, Golden Eagle, and Little Brown Crane were found.

Sayles, E. B.
1945. The San Simon Branch. Excavations at Cave Creek and in the San Simon Valley. I. Culture. Medallion Papers 34: 1–78.
Bones of Golden Eagle were found.

Schroeder, A. H.
1961. The archeological excavations at Willow Beach, Arizona. 1950. Univ. Utah Anthro. Papers 50: 1–163.
Lists bones of *Anas, Aechmophorus,* and *Zenaidura macroura.*

———
1961. Puerco ruin excavations, Petrified Forest National Monument, Arizona. Plateau 33: 93–104.
Remains of Eared Grebe found.

———
1968. Birds and feathers in documents relating to Indians of the southwest. Pp. 95–114, [in] A. H. Schroeder, Collected papers in honor of Lyndon Lane Hargrave. Papers Archeol. Soc. New Mex. 1: 1–170.
A summary with bibliography of 43 titles.

Skinner, M. F.
1942. The fauna of Papago Springs cave, Arizona, and a study of *Stockoceros;* with three new antilocaprines from Nebraska and Arizona. Bull. Amer. Mus. Nat. Hist. 80: 143–220.
Includes bones of an ibis-like bird, genus not determined.

Skinner, S. A.
1967. Four historic sites near Flagstaff, Arizona. Plateau 39: 105–123.
Brief mention of bones of the Mourning Dove at one of the sites.

Smith, W.
1952. Excavations in Big Hawk Valley, Wupatki National Monument, Arizona. Mus. N. Ariz. Bull. 24: 1–203.
Bones of Sparrow Hawk, Red-tailed Hawk, American Raven, and Roadrunner were found.

Tuthill, C.
1947. The Tres Alamos site on the San Pedro River, southeastern Arizona. Amerind Found. Inc., Dragoon, Ariz. vol. 4: 1–88.
Lists bones of Golden Eagle.

Wendorf, F.
1953. Archeological studies in the Petrified Forest National Monument. Mus. N. Ariz. Bull. 27: 1–203.
Mentions finding Turkey and Eagle bones.

Wetmore, A.
1924. Fossil birds from southeastern Arizona. Proc. U.S. Nat. Mus. 64 (2495): 1–18.
In the upper San Pedro Valley.

Wetmore, A.
1928. Birds of the past in North America. Smiths. Rep. App. 1928: 377–389.
Mentions Upper Pliocene fossils from Arizona.

———
1928. Prehistoric ornithology in North America. Jour. Wash. Acad. Sci. 18: 145–158.
Same as in Smithsonian Report Appendix of 1928.

———
1931. Early record of birds in Arizona and New Mexico. Condor 33: 35.
Parrots and Turkeys are mentioned in an account of the Espejo expedition of 1582–1583.

———
1940. A check-list of the fossil birds of North America. Smiths. Misc. Coll. 99 (4): 1–81.
Contains several Arizona records.

———
1943. Remains of a Swan from the Miocene of Arizona. Condor 45: 120.
From the vicinity of Wikieup.

———
1956. A check-list of the fossil and prehistoric birds of North America and the West Indies. Smiths. Misc. Coll. 131 (5): 1–105.

———
1957. A fossil Rail from the Pliocene of Arizona. Condor 59: 267–268.
Describes *Rallus phillipsi* from near Wikieup.

Wheat, J. B.
1954. Crooked Ridge Village. Univ. Ariz. Soc. Sci. Bull. 24, vol. 25 (3): 1–183.
Turkeys were used by the Indians on the San Carlos Apache Reservation.

———
1955. Mogollon culture prior to A.D. 1000. Amer. Anthro. Assoc. Mem. 82, vol. 57, no. 2, part 3: 1–242.
Mentions that Turkeys were hunted.

Wilson, R. W.
1942. Preliminary study of the fauna of Rampart Cave, Arizona. Carnegie Inst. Wash. Publ. 530: 169–185.
Some bird specimens found, but not yet identified.

Wyllys, R. K., ed.
1931. Padre Luís Velarde's Relación of Pimería Alta, 1716. New Mex. Hist. Rev. 6 (2): 111–157.
"Countless Turkeys in the mountains," page 129. At San Xavier del Bac many Macaws were raised for feathers by the Pimas.

4. Morphology and Physiology

Papers dealing with anatomy, molt, malformations, and physiology of Arizona birds are still few in number, but the emphasis on physiology of desert birds is increasing.

Austin, G. T.
1970. Experimental hypothermia in a Lesser Nighthawk. Auk 87: 372–374.
Includes comments on Marshall's 1955 study of Lesser Nighthawks.

——— and W. G. Bradley
1969. Additional responses of the Poor-will to low temperatures. Auk 86: 717–725.
A Mohave County, Arizona, bird was caught for the study.

Baker, B. and E. Baker
1952. Loggerhead Shrike with malformed bill. Wilson Bull. 64: 161.
Near Casa Grande, Arizona.

Brauner, J.
1952. Reactions of Poor-wills to light and temperature. Condor 54: 152–159.
Latest and earliest activity dates in Tucson were December 1, 1949, and February 4, 1950.

Breninger, G. F.
1897. An albino Green-tailed Towhee. Osprey 1: 137.
At Phoenix, Arizona.

Calder, W. A. and K. Schmidt-Nielsen
1967. Temperature regulation and evaporation in the Pigeon and the Roadrunner. Amer. Jour. Physiol. 213: 883–889.
The Roadrunners came from Arizona.

Carothers, S. W. and R. P. Balda
1970. Abnormal bill of a Western Meadowlark *Sturnella n. neglecta*. Auk 87: 173–174.
In Yavapai County, Arizona.

Chapin, J. P.
1921. The abbreviated inner primaries of nestling woodpeckers. Auk 38: 531–552.
Describes inner primaries of *Balanosphyra formicivora aculeata* from Pinal County, Arizona.

Clark, H. L.
1899. The feather-tracts of North American Grouse and Quail. Proc. U.S. Nat. Mus. 21: 641–653.
Based partly on specimens of *Colinus, Lophortyx,* and *Callipepla* from Arizona.

———
1906. The feather tracts of Swifts and Hummingbirds. Auk 23: 68–91.
Remarks on several species of Arizona Hummingbirds.

Cole, G. A.
1969. Plumage colors and patterns in the feral rock pigeons of central Arizona. Amer. Midl. Nat. 82: 613–618.
In Tempe, Arizona.

Dawson, W. R. and G. A. Bartholomew
1968. Temperature regulation and water economy of desert birds, pp. 357–394 [in] G. W. Brown, Jr. (ed.), Desert Biology. Academic Press, New York: 1–635.
Rufous-winged Sparrows in Arizona obtain water from insects (p. 384).

Foster, M. S.
1967. Molt cycles of the Orange-crowned Warbler. Condor 69: 169–200.
Considers as "questionable" the report that there is a complete postjuvenal molt in the Audubon and Myrtle Warbler (see Phillips, et al., 1964).

Hunt, M. J.
1959. Albino Gila Woodpecker. [In Letters in] Aud. Mag. 61: 195–196.
At Phoenix.

Johnson, N. K.
1963. Comparative molt cycles in the tyrannid genus *Empidonax*. [In] Proc. 13th Int. Ornith. Congr. 2: 870–883.
Arizona specimens studied, but localities not given.

Johnson, O. W.
1968. Some morphological features of avian kidneys. Auk 85: 216–228.
A number of the species were collected in Arizona.

Lasiewski, R. C. and R. J. Lasiewski
1967. Physiological responses of the Blue-throated and Rivoli's Hummingbirds. Auk 84: 34–48.
The Hummingbirds were captured in southeastern Arizona.

Lowe, C. H. and D. S. Hinds
1969. Thermoregulation in desert populations of Roadrunners and Doves, Abstract p. 113, [in] C. C. Hoff and M. L. Riedesel, Physiological systems in semiarid environments. Univ. New Mexico Press, Albuquerque.
Shading by adults is necessary to survival of nestlings.

MacMillen, R. E. and C. H. Trost
1965. Oxygen consumption and water loss in the Inca Dove, *Scardafella inca.* Am. Zool. 5: 208–209. Abstract of paper read at meeting.
Presumed to be Arizona.

——— and C. H. Trost
1966. Water economy and salt balance in White-winged and Inca Doves. Auk 83: 441–456.
Seventeen White-winged Doves and 27 Inca Doves were trapped in or near Tucson, Arizona, and shipped to California for study.

——— and C. H. Trost
1967. Thermoregulation and water loss in the Inca Dove. Comp. Biochem. Physiol. 20: 263–273.
Tucson birds were used.

——— and C. H. Trost
1967. Nocturnal hypothermia in the Inca Dove, *Scardafella inca.* Comp. Biochem. Physiol. 23: 243–253.
Some notes on roosting behavior at Tucson.

Marler, P.
1967. Comparative study of song development in sparrows. [In] Proc. 14th Int. Ornith. Congr.: 231–244.
Arizona Junco studied, but locality not given.

Marshall, J. T., Jr.
1955. Hibernation in captive Goatsuckers. Condor 57: 129–134.
A report on experiments on Nighthawks and Poor-wills at Tucson, Arizona.

Miller, A. H.
1933. Postjuvenal molt and the appearance of sexual characters of plumage in *Phainopepla nitens.* Univ. Calif. Publ. Zool. 38: 425–445.
Based partly on some Arizona specimens.

[Miller, A. H.]
1948. [Note on Allan Brooks' colored frontispiece, opposite page 137, of the Coppery-tailed Trogon in the Santa Rita Mountains, Arizona.] [In] Notes and News, Condor 50: 166.
Original color notes are given.

Neal, B. J.
1965. Notes on hematocrits from Arizona mammals and birds. S. W. Nat. 10: 69–72.
The birds listed are: English Sparrow, Mourning Dove, and Red-winged Blackbird.

Ohmart, R. D. and E. L. Smith
1970. Use of sodium chloride solutions by the Brewer's Sparrow and Tree Sparrow. Auk 87: 329–341.
Brewer's Sparrows from the Tucson area had low water requirements.

Phillips, A. R.
1948. Survival of birds at high temperatures. Amer. Nat. 82: 331–334.
Arizona temperatures are not lethal to English Sparrows.

Phillips, A. R.
1951. The molts of the Rufous-winged Sparrow. Wilson Bull. 63: 323–326.
Includes data on Arizona specimens.

———
1966. Some unusual Vireos. Bird-Banding 37: 286–287.
Discusses albinos collected in Arizona.

Pitelka, F. A.
1945. Pterylography, molt, and age determination of American Jays of the genus *Aphelocoma*. Condor 47: 229–260.
An extensive report that includes Arizona birds.

Rand, A. L.
1961. Wing length as an indicator of weight: a contribution. Bird-Banding 32: 71–79.
Included are measurements and weights of Scrub and Mexican Jays from Arizona.

Rea, A. M.
1967. Age determination of *Corvidae*. Part 1: Common Crow. Western Bird Bander 42: 44–47.
Includes figures of Arizona specimens of *Corvus brachyrhynchos hargravei*.

——— and D. Kanteena
1968. Age determination of *Corvidae*. Part II: Common and White-necked Ravens. Western Bird Bander 43: 6–9.
Includes figures of Arizona specimens.

Ricklefs, R. E.
1968. Patterns of growth in birds. Ibis 110: 419–451.
Includes data on Cactus Wrens and Curve-billed Thrashers in Arizona.

——— and F. R. Hainsworth
1968. Temperature dependent behavior of the Cactus Wren. Ecol. 49: 227–233.

——— and F. R. Hainsworth
1968. Temperature regulation in nestling Cactus Wrens: the development of homeothermy. Condor 70: 121–127.

——— and F. R. Hainsworth
1969. Temperature regulation in nestling Cactus Wrens: the nest environment. Condor 71: 32–37.

Ridgway, R.
1872. On the relation between color and geographical distribution in birds, as exhibited in melanism and hyperchromism. Amer. Jour. Sci. 4 (24), art. 55: 454–460.
Mentions coloration of Arizona specimens of Goldfinches.

———
1873. On the relation between color and geographical distribution in birds, as exhibited in melanism and hyperchromism. Amer. Jour. Sci. 5 (25), art. 4: 39–44.
Mentions variations in color of Cardinals and Steller Jays in Arizona.

Ross, C. C.
1963. Albinism among North American birds (annotated list). Cassinia 47: 2–21.

Russell, S. M.
1969. Regulation of egg temperatures by incubating White-winged Doves, pp. 107–112, [in] C. C. Hoff and M. L. Riedesel, eds., Physiological systems in semiarid environments. Univ. New Mexico Press, Albuquerque.
Under high ambient temperatures, eggs are cooled by heat transfer to the brood patch.

Saiza, A.
1968. Age determination of Corvidae. Part III. Juvenals. Western Bird Bander 43: 20–23.
Includes figures of Arizona specimens.

Salt, G. W.
1963. Avian body weight, adaptation, and evolution in western North America. [In] Proc. 13th Int. Ornith. Cong. 2: 905–917.
Populations vary geographically in body weight. Some Arizona species included, with maps.

Smith, A. P.
1908. Albinism of Scaled Partridge. Condor 10: 93.
Two specimens of *Callipepla squamata* from the San Pedro Valley, Arizona.

Smith, E. L. and R. D. Ohmart
1969. Water economy of the Green-tailed Towhee *(Chlorura chlorura)*. Pp. 115–124, [in] C. C. Hoff and M. L. Riedesel, eds., Physiological systems in semiarid environments. Univ. New Mexico Press, Albuquerque.
They could not use salt water as drinking water.

Stephens, F.
1904. Cactus Wrens. Condor 6: 51–52.
Remarks on the coloration of Arizona specimens.

Thornburg, F.
1953. Another hibernating Poor-will. Condor 55: 274.
In the Silverbell Mountains, Arizona.

5. Ecology

Many entries here reveal the extent and importance of investigations in this field. Included here are descriptions of habitat, nesting, eggs, food, and interactions with other species of animals in the environment.

Adios
1884. Some Arizona Quails. Forest and Stream 22: 103.
Indians snare Quail by the thousands.

Alcorn, S. M., S. E. McGregor, and G. Olin
1961. Pollination of saguaro cactus by Doves, nectar-feeding bats, and honey bees. Science 133 (3464): 1594–1595.
White-winged Doves were effective in cross-pollination.

Aldous, S. E.
1942. The White-necked Raven in relation to agriculture. U.S. Dept. Int. Fish and Wildl. Serv. Research Rep. 5: 1–56.
Includes Arizona distribution and nesting.

Allen, R. W. and M. M. Nice
1952. A study of the breeding biology of the Purple Martin *(Progne subis)*. Amer. Midland Nat. 47: 606–665.
Includes observations on nesting in Arizona by M. B. Cater.

Amadon, D.
1940. Hosts of the Cowbirds. Auk 57: 257.
Cooper Tanager at Tucson parasitized by the Bronzed Cowbird.

——— and A. R. Phillips
1939. Notes on the Mexican Goshawk. Auk 56: 183–184.
In the Santa Cruz River bottoms south of Tucson; description of nest, male nestling, and food.

Anderson, A. H.
1933. A tail-less Roadrunner. Bird-Lore 35: 95–96.
Brief notes on a nest at Tucson.

———
1933. Electrocution of Purple Martins. Condor 35: 204.
Near Tucson.

———
1934. Food of the Gila Woodpecker *(Centurus uropygialis uropygialis).* Auk 51: 84–85.
It feeds on gall insects at Tucson.

———
1934. Cactus Wrens and Thrashers. Bird-Lore 36: 108–109.
Remarks on nest destruction by Thrashers and the hazards of thorny shrubs.

———
1934. Notes on the Rock Wren. Bird-Lore 36: 173–174.
At Tucson.

———
1934. A Cactus Wren roosting in a Verdin's nest. Bird-Lore 36: 366.
At Tucson.

———
1965. Notes on the behavior of the Rufous-winged Sparrow. Condor 67: 188–190.
At Tucson.

——— and A. Anderson
1944. 'Courtship feeding' by the House Finch. Auk 61: 477–478.
At Tucson.

——— and A. Anderson
1946. Notes on the Purple Martin roost at Tucson, Arizona. Condor 48: 140–141.

——— and A. Anderson
1946. Notes on the use of creosote bush by birds. Condor 48: 179.

——— and A. Anderson
1946. Late nesting of the Pyrrhuloxia at Tucson, Arizona. Condor 48: 246.

——— and A. Anderson
1946. Notes on the Cedar Waxwing at Tucson, Arizona. Condor 48: 279–280.
Feeding habits.

——— and A. Anderson
1948. Observations on the Inca Dove at Tucson, Arizona. Condor 50: 152–154.

——— and A. Anderson
1948. Notes on two nests of the Beardless Flycatcher near Tucson, Arizona. Condor 50: 163–164.

——— and A. Anderson
1957. Life history of the Cactus Wren. Part I: winter and pre-nesting behavior. Condor 59: 274–296.
At Tucson, Arizona.

——— and A. Anderson
1959. Life history of the Cactus Wren. Part II: The beginning of nesting. Condor 61: 186–205.

Anderson, A. H. and A. Anderson
1960. Life history of the Cactus Wren. Part III: The nesting cycle. Condor 62: 351–369.

——— and A. Anderson
1961. Life history of the Cactus Wren. Part IV: Development of nestlings. Condor 63: 87–94.

——— and A. Anderson
1962. Life history of the Cactus Wren. Part V: Fledging to independence. Condor 64: 199–212.

——— and A. Anderson
1963. Life history of the Cactus Wren. Part VI: Competition and Survival. Condor 65: 29–43.

——— and A. Anderson
1965. The Cactus Wrens on the Santa Rita Experimental Range, Arizona. Condor 67: 344–351.
A population study.

Anonymous
1926. Suicide Robin returns. Grand Canyon Nature Notes 1 (1): 4.
Flies into window.

———
1926. Any port in a storm. Grand Canyon Nature Notes 1 (3): 5.
A Thurber Junco seeks shelter on a porch during a snow storm.

———
1926. Intelligent Robin. Grand Canyon Nature Notes 1 (5): 4.
Robins dipped dry pine needles in water at drinking fountain to make them pliable for nest construction!

———
1927. Mother Grouse poses for movie camera. Grand Canyon Nature Notes 2 (4): 3–4.
Dusky Grouse on North Rim of the Grand Canyon.

———
1928. Robin breaks ice to take bath. Grand Canyon Nature Notes 3 (4): 2.

———
1929. Nesting notes. Grand Canyon Nature Notes 3 (10): 3.
From records of Mrs. F. M. Bailey and the park naturalist.

Antevs, A.
1947. Cactus Wrens use "extra" nest. Condor 49: 42.
Near Globe, Arizona.

———
1947. Towhee helps Cardinal feed their fledglings. Condor 49: 209.
At Globe, Arizona.

———
1948. Behavior of the Gila Woodpecker, Ruby-crowned Kinglet, and Broad-tailed Hummingbird. Condor 50: 91–92.
At Globe, Arizona.

Anthony, A. W.
1891. Notes on the Cactus Wren. Zoe 2: 133–134.
Discusses nesting habits in California, Arizona, and New Mexico.

Arnold, L.
1941. The mesquite forest and the Whitewing. Ariz. Wildlife and Sportsman 3 (11): 5–6.
Some early history of the Santa Cruz Valley.

Arnold, L. W.
1942. Water birds influenced by irrigation projects in the lower Colorado River Valley. Condor 44: 183–184.
Remarks on 20 species.

———
1942. The aerial capture of a White-throated Swift by a pair of Falcons. Condor 44: 280.
In the vicinity of the Castle Dome Mountains, Arizona.

———
1943. The Western White-winged Dove in Arizona. Pittman-Robertson Project Arizona 9-R, Arizona Game and Fish Commission, Federal Aid Division. 103 pp.

———
1954. The Golden Eagle and its economic status. U.S. Dept. Int. Fish and Wildl. Serv. Circ. 27: 1–35.
Brief notes on its prey in Arizona.

Bailey, F. M.
1922. Koo. Bird-Lore 24: 260–265.
An account of a Roadrunner as observed in southern Arizona.

———
1922. Cactus Wrens' nests in southern Arizona. Condor 24: 163–168.
An important, detailed account of roosting nests, their sites and construction, at the north base of the Santa Rita Mountains.

———
1923. Fifteen Arizona Verdins' nests. Condor 25: 20–21.
Account of roosting nests found at the foot of the Santa Rita Mountains.

———
1932. Abert Squirrel burying pine cones. Jour. Mamm. 13: 165–166.
Includes a report of a Mearn's Woodpecker chasing an Abert Squirrel at Grand Canyon.

———
1940. Birds in southwestern Indian life. Kiva 5 (5): 17–20.
A good summary; includes Arizona, Utah, and New Mexico.

Balda, R. P.
1965. Loggerhead Shrike kills Mourning Dove. Condor 67: 359.
In southeastern Arizona.

———
1968. Foliage use by birds of the oak-juniper-pine woodland in southeastern Arizona. Abstract of paper read at 12th annual meeting of the Ariz. Acad. Science at Flagstaff. Jour. Ariz. Acad. Sci. 5, proc. suppl: 15–16.

Balda, R. P.
1969. Foliage use by birds of the oak-juniper woodland and ponderosa pine forest in southeastern Arizona. Condor 71: 399–412.
Oak-juniper woodland: 36 species of birds, 267 pairs per 100 acres; pine forest: 31 species, 336 pairs per 100 acres.

———
1970. Effects of spring leaf-fall on composition and density of breeding birds in two southern Arizona woodlands. Condor 72: 325–331.
In the Chiricahua Mountains: species density greater in oak-juniper-pine woodland.

Bancroft, G.
1946. Geographic variation in the eggs of Cactus Wrens in Lower California. Condor 48: 124–128.
Measurements of Arizona specimens are included.

Barlow, J. C., R. D. James, and N. Williams
1970. Habitat co-occupancy among some Vireos of the subgenus *Vireo* (Aves: Vireonidae). Canadian Jour. Zool. 48: 395–398.
In central and northern Arizona.

[Barnes, R. M.]
1915. Abbott's collection of North American Warblers' eggs. Ool. 32: 129–130.
Eggs of several Arizona species are listed.

Baumgartner, F. M.
1938. Courtship and nesting of the Great Horned Owls. Wilson Bull. 50: 274–285.
Egg-laying in Arizona occurs from late February to early April.

Beal, F. E. L.
1911. Food of the Woodpeckers of the United States. U.S. Dept. Agric. Biol. Surv. Bull. 37: 1–64.
Brief account of stomach contents of the Gila Woodpecker and Gilded Flicker.

Beaton, E. S.
1927. Road-runners seize marauding snake. Ariz. Wild Life 1 (2): 11–12.
In the Salt River Valley.

Beecher, W. J.
1950. Convergent evolution in the American Orioles. Wilson Bull. 62: 51–86.
A broad, detailed discussion that includes references to Arizona birds and environment.

Beidleman, R. G.
1956. Ethnozoology of the Pueblo Indians in historic times. Part II: Birds. Southwestern Lore 22: 17–28.
A summary with list of references.

Bendire, C. E.
1873. Nest, eggs and breeding habits of the Vermilion Flycatcher *(Pyrocephalus rubineus* var. *Mexicanus).* Amer. Nat. 7: 170–171.
In southern Arizona.

———

1878. Breeding habits of *Geococcyx californianus*. Bull. Nuttall Ornith. Club 3: 39.

Twenty nests found in southern Arizona in 1872.

———

1890. Notes on *Pipilo fuscus mesoleucus* and *Pipilo aberti*, their habits, nests and eggs. Auk 7: 22–29.

———

1892. Life histories of North American birds with special reference to their breeding habits and eggs. U.S. Nat. Mus. Special Bull. 1: 1–446.

———

1893. The Cowbirds. U.S. Nat. Mus. Rep. 1893: 589–624.

Includes some records of parasitism of Arizona species.

———

1895. Life histories of North American birds, from the parrots to the grackles, with special reference to their breeding habits and eggs. U.S. Nat. Mus. Special Bull. 3: 1–518.

Bené, F.

1940. Rhythm in the brooding and feeding routine of the Black-chinned Hummingbird. Condor 42: 207–212.

Studies made at Phoenix, Arizona.

———

1941. Experiments on the color preference of Black-chinned Hummingbirds. Condor 43: 237–242.

At Phoenix, Arizona, red did not attract the birds more than any other color.

———

1945. The role of learning in the feeding behavior of Black-chinned Hummingbirds. Condor 47: 3–22.

Observations at Phoenix, Arizona.

———

[1947] The feeding and related behavior of hummingbirds with special reference to the Black-chin, *Archilochus alexandri* (Bourcier and Mulsant). Mem. Boston Soc. Nat. Hist. 9: 395–478.

At Phoenix, Arizona.

Bent, A. C.

1919. Life histories of North American diving birds. U.S. Nat. Mus. Bull. 107: 1–245. 1921. Bull. 113: 1–345, Gulls and Terns. 1922. Bull. 121: 1–343, orders Tubinares and Steganopodes. 1923. Bull. 126, part [1]: 1–245, order Anseres. 1925. Bull. 130, part [2]: 1–376, order Anseres. 1926. Bull. 135: 1–490, orders Odontoglossae, Herodiones, and Paludicolae. 1927. Bull. 142: 1–420, order Limicolae (part 1). 1929. Bull. 146: 1–412, order Limicolae (part 2). 1932. Bull. 162: 1–490, orders Galliformes and Columbiformes. 1937. Bull. 167: 1–409, order Falconiformes (part 1). 1938. Bull. 170: 1–482, orders Falconiformes and Strigiformes (part 2). 1939. Bull. 174: 1–334, order Piciformes. 1940. Bull. 176: 1–506, Cuckoos, Goatsuckers, Hummingbirds and their allies. 1942. Bull. 179: 1–555, order Passeriformes: Flycatchers,

Larks, Swallows, and their allies. 1946. Bull. 191: 1–495, Jays, Crows, and Titmice. 1948. Bull. 195: 1–475, Nuthatches, Wrens, Thrashers and their allies. 1949. Bull. 196: 1–454, Thrushes, Kinglets, and their allies. 1950. Bull. 197: 1–411, Wagtails, Shrikes, Vireos, and their allies. 1953. Bull. 203: 1–734, Wood Warblers. 1958. Bull. 211: 1–549, Blackbirds, Orioles, Tanagers, and allies.

——— and collaborators. O. L. Austin, Jr., ed.
1968. Life histories of North American Cardinals, Grosbeaks, Buntings, Towhees, Finches, Sparrows and allies. Order Passeriformes: Family Fringillidae. U.S. Nat. Mus. Bull. 237, part 1: 1–602; part 2: 603–1248; part 3: 1249–1889.

Birchett, J. T.
1948. [Notes on the Gila Woodpecker, House Finch, and Inca Dove at Phoenix.] [In Letters in] News from the Bird-Banders 23: 22.

Blackford, J. L.
1953. Breeding haunts of the Stephens Whip-poor-will. Condor 55: 281–286.
Photographs of nest and eggs and incubating bird, in the Chiricahua Mountains, Arizona.

Bodensten, A. J.
1939. Old time collectors. Ool. 56: 26–31.
Lists "Palmers Thrasher; Tuscon [sic], Ariz.; May 21, 1911; 3 eggs; F. C. Willard."

Borell, A. E.
1936. A modern La Brea tar pit. Auk 53: 298–300.
List of birds found dead in a man-made tar pit at the south rim of the Grand Canyon.

———
1937. Cooper Hawk eats a Flammulated Screech Owl. Condor 39: 44.
At Grand Canyon National Park.

Brandt, H.
1937. Some Arizona bird studies. Auk 54: 62–64.
Observations on 6 species in the Huachuca Mountains.

Breninger, G. F.
1897. A roosting method of the Inca Dove. Osprey 1: 111.
At Phoenix, Arizona.

———
1897. An unusual nesting site. Osprey 1: 122.
A Gambel Quail nests in a hollow of a mesquite tree.

———
1897. Coues' Flycatcher. Osprey 2: 12.
A nest in the Huachuca Mountains.

———
1897. Nocturnal flights of the Turkey Vulture. Osprey 2: 54–55.
Near Tucson.

———
1899. Gambel's Quail. Osprey 3: 84–85.
Observations in southern Arizona.

———
1899. A nest of the Blue-throated Hummingbird. Osprey 3: 86.
In the Huachuca Mountains.

———
1905. Are the habits of birds changing? Auk 22: 360–363.
Unusual nesting sites of several species in southern Arizona.

Brewer, T. M.
1873. Description of some nests and eggs of Arizona birds. Proc. Boston Soc. Nat. Hist. 16: 106–111.
Fourteen species from Bendire's collection near Tucson.

———
1878. Notes on *Junco caniceps* and the closely allied forms. Bull. Nuttall Ornith. Club 3: 72–75.
Describes eggs of *J. cinereus* and *J. dorsalis* from Arizona.

———
1879. Notes on the nests and eggs of the eight North American species of *Empidonaces*. Proc. U.S. Nat. Mus. 2: 1–10.
Includes *E. obscurus* from Arizona.

———
1879. The Cow-blackbird of Texas and Arizona *(Molothrus obscurus)*. Bull. Nuttall Ornith. Club 4: 123.
Includes measurements of a set of eggs from Arizona.

Brewster, W.
1882. Nest and eggs of *Setophaga picta*—a correction. Bull. Nuttall Ornith. Club 7: 249.
Corrected to third clutch instead of first.

Bristow, B.
1969. The fatal future. Part V. Statewide summary. Wildlife Views 16 (5): 13–21.
The clearing of phreatophytes endangers Arizona wildlife.

Brooks, S. C.
1943. Speed of flight of Mourning Doves. Condor 45: 119.
East of the Painted Desert two doves flew at a speed of 43–45 miles per hour.

Brown, H.
1885. Arizona bird notes. Forest and Stream 24: 367.
Nesting notes on Thrashers, Owls, and Gambel's Quail at Tucson and Quihotoa.

———
1885. Arizona Quail notes. Forest and Stream 25: 445.
Chiefly on *Colinus ridgwayi*.

———
1885. Peculiar eggs of *Scops trichopsis*. Ornith. and Ool. 10: 96.
Evidently *Otus asio*. He speculates that there is hybridism between that species and *Falco sparverius!*

Brown, H.
1887. Arizona bird notes. Forest and Stream 27: 464.

———

1888. On the nesting of Palmer's Thrasher. Auk 5: 116–118.
Near Tucson, Arizona.

———

1892. The habits and nesting of Palmer's Thrasher. Zoe 3: 243–248.
In southern Arizona.

———

1900. The conditions governing bird life in Arizona. Auk 17: 31–34.
Drought and overgrazing are discussed.

———

1901. Bendire's Thrasher. Auk 18: 225–231.
Describes the habitat, nests, and eggs.

———

1901. A Band-tailed Hawk's nest—an Arizona incident of biographical interest. Auk 18: 392–393.
Nest of *Buteo abbreviatus* east of Tucson.

———

1903. Arizona bird notes. Auk 20: 43–50.
In the vicinity of Yuma.

———

1904. The Elf Owl in California. Condor 6: 45–47.
Brief account of nesting habits in Arizona.

Brown, J. L.
1963. Social organization and behavior of the Mexican Jay. Condor 65: 126–153.
Observations in the Santa Rita Mountains.

Bryant, H. C.
1943. Birds eat snow. Condor 45: 77.
Juncos, Mountain Chickadees, and English Sparrows ate snow in the Grand Canyon National Park.

Bryant, W. E.
1881. Nest and eggs of the Painted Flycatcher *(Setophaga picta).* Bull. Nuttall Ornith. Club 6: 176–177.
From the Santa Rita Mountains.

Bull, M. T.
1948. [Behavior of an Arizona Cardinal in traps.] [In Letters with the annual reports]. News from the Bird-Banders 23: 33.

Calder, W. A.
1967. Breeding behavior of the Roadrunner, *Geococcyx californianus.* Auk 84: 597–598.
The birds studied were reared in Arizona.

———

1968. The diurnal activity of the Roadrunner, *Geococcyx californianus.* Condor 70: 84–85.
On the Santa Rita Experimental Range activity decreased at midday.

Calhoun, J. B.
1947. The role of temperature and natural selection in relation to the variations in the size of the English Sparrow in the United States. Amer. Nat. 81: 203–228.
Arizona specimens are mentioned.

Campbell, B.
1934. Report on a collection of reptiles and amphibians made in Arizona during the summer of 1933. Occas. Papers Mus. Zool. Univ. Mich. 12 (289): 1–10.
An 8-inch ring snake *(Diadelphis regalis arizonae)* was found in the stomach of a juvenile *Otus asio cineraceus*.

Cardwell, L.
1948. *Meleagris gallopavo merriami*. Ariz. Highways 24 (10): 38–39.
Brief notes on habits; one photograph.

Carothers, S. W.
1968. The relationship between diurnal and nocturnal respiratory metabolism and temperature in two birds of the genus *Junco*. Abstract of paper read at 12th annual meeting of the Ariz. Acad. Science at Flagstaff. Jour. Ariz. Acad. Sci. 5, proc. suppl.: 16.

Carr, W. H.
1945. Watching the White-winged Dove. Nat. Hist. 54: 412–417.
Nesting notes from vicinity of Tucson.

———
1947. Trailing desert owls. Nat. Hist. 56: 468–473.
Notes on nesting of Great Horned Owls in southern Arizona.

Castetter, E. F. and R. M. Underhill
1935. The ethnobiology of the Papago Indians. Univ. New Mex. Ethnobiological studies in the American southwest II. Univ. New Mex. Bull. Biol. ser. 4 (3): 1–84.
Lists birds used for food and other birds caged for feathers.

Cater, M. B.
1944. Roosting habits of Martins at Tucson, Arizona. Condor 46: 15–18.

Chapel, W. L.
1939. Wildlife vs. logging. Ariz. Wildlife and Sportsman 1 (1): 9.
Turkey nesting in logged areas near Williams.

[Childs, J. L.]
1905. Eggs of the Olive Warbler *(Dendroica olivacea)*. The Warbler, 2nd ser. 1: 17; pl. I.
From the Huachuca Mountains.

———
1906. Nest and eggs of the Blue-throated Hummingbird *(Coeligena clemenciae)*. The Warbler 2: 65; pl. IV.
From the Huachuca Mountains.

Clark, J. H.
1900. The giant cactus as a nesting place for the Western Red-tailed Hawk. Ool. 17: 126.
Descriptions of six nests and their eggs from the region "south of Tucson, Arizona."

Cole, G. A. and M. C. Whiteside
1965. Kiatuthlanna—a limnological appraisal. II. Chemical factors and biota. Plateau 38: 36–48.
In Apache County, Arizona; a Coot found dead in the salt pond.

Collins, G. L.
1934. [Shufeldt Junco.] Grand Canyon Nature Notes 8: 237.
A banded Junco flew into a window and was killed.

Colton, H. S.
1939. Prehistoric culture units and their relationships in northern Arizona. Mus. N. Ariz. Bull. 17: 1–76.
Turkey is listed as a food item in several culture units.

———
1945. An unusual accident to a Broad-tailed Hummingbird. Plateau 18: 15.
At Flagstaff, Arizona.

———
1960. Black sand. Prehistory in northern Arizona. Univ. New Mex. Press: 1–132.
The Hohokam traded Macaws to the northern Indian tribes (p. 88).

Cook, R. E.
1969. Variation in species density of North American birds. Syst. Zool. 18: 63–84.
High density in Arizona.

Cooke, M. T.
1937. Flight speed of birds. U.S. Dept. Agric. Circ. 428: 1–14.
Quoting Grinnell, a Pintail in Arizona flew at a speed of 52+ miles per hour.

Cottam, C.
1947. Zone-tailed Hawk feeds on rock squirrel. Condor 49: 210.
Near Globe, Arizona.

——— and P. Knappen
1939. Food of some uncommon North American birds. Auk 56: 138–169.
Food of 11 Arizona species is listed, including that of the Masked Bob-white.

———, C. S. Williams, and C. A. Sooter
1942. Flight and running speeds of birds. Wilson Bull. 54: 121–131.
Gambel's Quail, 11 and 14 miles per hour; Roadrunner, 12 miles per hour in Arizona.

Coues, E.
1872. The nest, eggs, and breeding habits of *Harporhynchus crissalis.* Amer. Nat. 6: 370–371.
Along Rillito Creek, near Tucson.

———
1872. Nest and eggs of *Helminthophaga Luciae.* Amer. Nat. 6: 493.
Near Tucson.

———
1882. Nesting of the White-bellied Wren *(Thryothorus bewicki leucogaster).* Bull. Nuttall Ornith. Club 7: 52–53.
In northern Arizona.

Crockett, H. L. and R. Crockett
1936. A tree nesting Quail. Condor 38: 97–99.
Gambel's Quail at Phoenix.

Crockett, R. M.
1936. Shrike craftiness. Condor 38: 88.
Shrike imitates House Finch calls at Phoenix.

Dale, E. E.
1949. The Indians of the southwest. Univ. Okla. Press, Norman. 283 pp.
Describes capture of Eagles used in rain ceremonies.

Davie, O.
1889. Nests and eggs of North American birds. Hann and Adair, Columbus. 4th ed. 455 pp.

Davis, L. I.
1961. Songs of North American *Myiarchus*. Texas Jour. Sci. 13: 327–344.
Myiarchus cinerescens [sic] at Blythe, Arizona [sic].

De Benedictis, P.
1966. The flight song display of two taxa of Vermilion Flycatcher, genus *Pyrocephalus*. Condor 68: 306–307.
Includes observations on *P. rubinus* from the lower Colorado River Valley.

deLaubenfels, M. W.
1925. Unusual notes of Texas Nighthawk. Condor 27: 210.
Near Phoenix.

Demaree, S. R.
1970. Nest-building, incubation period, and fledging in the Black-chinned Hummingbird. Wilson Bull. 82: 225.
At Phoenix, Arizona.

Dille, F. M.
1936. Arizona days. Ool. 53: 67.
Comments on Shrikes.

———
1940. Arizona days. Ool. 57: 86–87.
Brief notes on Cowbirds.

Dixon, K. L.
1950. Notes on the ecological distribution of Plain and Bridled Titmice in Arizona. Condor 52: 140–141.

———
1959. Ecological and distributional relations of desert scrub birds of western Texas. Condor 61: 397–409.
Includes comparisons with Arizona habitats.

———
1961. Habitat distribution and niche relationships in North American species of *Parus*. Pp. 179–216, [in] W. F. Blair, ed., Vertebrate speciation, a University of Texas Symposium. Univ. Texas Press, Austin.
All Arizona species of *Parus* are discussed.

Dodge, H. H.
1894. Dove life in Arizona. Ool. 11: 229–230.
Notes on their habits at Phoenix.

Dorsey, G. A.
1903. Indians of the southwest. A. T. and S. F. Railway System. 223 pp.
Mentions Eagle and Turkey feathers.

Eaton, W. C.
1933. Frontier life in southern Arizona, 1858–1861. SW'n Hist. Quarterly 36: 173–192.
Mentions Swallows nesting in the San Xavier Church south of Tucson.

Elder, J. B.
1956. Watering patterns of some desert game animals. Jour. Wildl. Mgmt. 20: 368–378.
Data on Doves and Quail in the Tucson Mountain Park.

Engels, W.
1940. Structural adaptations in Thrashers (Mimidae: genus *Toxostoma*) with comments on interspecific relationships. Univ. Calif. Publ. Zool. 42: 341–400.
Includes the results of considerable field work in the Santa Cruz Valley, Arizona.

Evermann, B. W.
1882. Black-crested Flycatcher. Ornith. and Ool. 7: 169–170, 177–179. *Phainopepla nitens*. Includes quotations from Bendire on nesting in Arizona.

Falvey, E. B.
1936. His majesty, the Mearn's Quail. Game Breeder and Sportsman 40: 226–227, 241.
Includes a brief account of nesting in southern Arizona.

Fewkes, J. W.
1900. Property rights in Eagles among the Hopi. Amer. Anthro. 2, new ser.: 690–707.
Eagles were properties of clans; hunting practices are described.

Field, C. L.
1935. A reply to H. H. Kimball. Ool. 52: 104.
Elf Owl lays from 3–5 eggs in a set. Four eggs at 4,500 feet in the Santa Catalina Mountains.

Finley, W. L. and I. Finley
1915. With the Arizona Roadrunners. Bird-Lore 17: 159–165.
Notes on habits, with excellent photographs from the Tucson region.

Fitch, F. W., Jr.
1950. Life history and ecology of the Scissor-tailed Flycatcher, *Muscivora forficata*. Auk 67: 145–168.
Lists two Arizona records.

Friedmann, H.
1929. The Cowbirds. A study in the biology of social parasitism. Charles C. Thomas, Springfield, Ill. 421 pp.
Lists a number of Arizona species that have been parasitized.

———

1931. Additions to the list of birds known to be parasitized by the Cowbirds. Auk 48: 52–65.

Dendroica graciae, a new host.

———

1933. Further notes on the birds parasitized by the Red-eyed Cowbird. Condor 35: 189–191.

Arizona Hooded Orioles at Tombstone and Oracle, Arizona, are listed as victims.

———

1934. Further additions to the list of birds victimized by the Cowbird. Wilson Bull. 46: 25–36.

Adds *Tyrannus vociferans, Pyrocephalus rubinus mexicanus, Toxostoma bendirei, Vireo belli arizonae,* and *Vireo solitarius plumbeus*, all from Arizona.

———

1934. Further additions to the list of birds victimized by the Cowbird. Wilson Bull. 46: 104–114.

Includes the Pyrrhuloxia, Rufous-winged Sparrow, and Song Sparrow from Arizona.

———

1943. Further additions to the list of birds known to be parasitized by the Cowbirds. Auk 60: 350–356.

Vermilion Flycatcher parasitized by the Dwarf Cowbird at Nogales, Arizona.

———

1963. Host relations of the parasitic Cowbirds. U.S. Nat. Mus. Bull. 233: 1–276.

Contains numerous references to Arizona species of birds parasitized by the Cowbirds.

———

1966. Additional data on the host relations of the parasitic Cowbirds. Smiths. Misc. Coll. 149: 1–12.

Contains several records from Arizona.

George, W. G.

1963. Phylogenetic riddle. Nat. Hist. 72: 45–47.

Discussion of the Olive Warbler, with good photo of nest with young; presumably in Arizona.

Gilman, M. F.

1909. Among the thrashers in Arizona. Condor 11: 49–54.

Observations on five species of thrashers in the Pima Indian Reservation.

———

1909. Some owls along the Gila River in Arizona. Condor 11: 145–150.

Observations on *Bubo virginianus, Aluco pratincola, Otus trichopsis* (= *Otus asio*), *Speotyto cunicularia, Glaucidium phalaenoides,* and *Micropallas whitneyi.*

———

1909. Nesting notes on the Lucy Warbler. Condor 11: 166–168.

Along the Gila River, Arizona.

Gilman, M. F.
1909. Red-eyed Cowbird at Sacaton, Arizona. Condor 11: 173.

———
1910. Notes from Sacaton, Arizona. Condor 12: 45–46.
Records of *Calcarius ornatus, Coturnicops noveboracensis, Baeolophus wollweberi.* The Red-eyed Cowbird is *T. a. aeneus.*

———
1911. Notes from Sacaton, Arizona. Condor 13: 35.
Observations on 7 species of birds.

———
1911. Doves on the Pima Reservation. Condor 13: 51–56.
Notes on *Zenaidura macroura, Melopelia asiatica, Chaemepelia passerina,* and *Scardafella inca.*

———
1915. Woodpeckers of the Arizona lowlands. Condor 17: 151–163.
Notes on 8 species along the Gila River.

Glover, F. A.
1953. Summer foods of the Burrowing Owl. Condor 55: 275.
In Maricopa County, Arizona.

Gorsuch, D. M.
1932. The Roadrunner. Ariz. Wild Life 4 (4): 1–2, 11.
Food items discussed.

———
1934. Life history of the Gambel Quail in Arizona. Univ. Ariz. Biol. Sci. Bull. 2: 1–89.

———
1936. Banding records of Gambel Quail. Condor 38: 126.
On the Santa Rita Experimental Range; two of the birds lived to be at least 5-1/2 years old.

Gould, P. J.
1961. Territorial relationships between Cardinals and Pyrrhuloxias. Condor 63: 246–256.
Observations in the San Xavier Indian Reservation near Tucson, Arizona.

Grant, K. A. and V. Grant
1967. Records of hummingbird pollination in the western American flora. III. Arizona records. Aliso 6: 107–110.
Lists several species in mountain areas.

——— and V. Grant
1968. Hummingbirds and their flowers. Columbia Univ. Press, New York and London. 115 pp. 30 pl.
Includes all Arizona breeding species.

Grant, P. R.
1966. The coexistence of two wren species of the genus *Thryothorus.* Wilson Bull. 78: 266–278.
Includes a brief discussion of the relationships of the Cardinal and Pyrrhuloxia and the Abert and Brown Towhees in Arizona.

Grater, R. K.
1934. [Chipping Sparrow attacks chipmunk.] Grand Canyon Nature Notes 9: 303.

———
1934. An archaeological minded Jay. Grand Canyon Nature Notes 9: 322–323.
Inquisitiveness in the Grand Canyon museum.

———
1934. Notes on the Band-tailed Pigeon. Grand Canyon Nature Notes 9: 330–331.
At Grand Canyon Village they came for water.

———
1934. [Accidents to a Chestnut-backed Bluebird.] Grand Canyon Nature Notes 9: 337.

———
1939. Further notes on the feeding habits of the Treganza Heron. Condor 41: 217.
Feeding on rodents in Lake Mead.

Grinnell, J.
1908. Goonies of the desert. Condor 10: 92.
Ravens observed following trains.

Gross, A. O.
1949. Nesting of the Mexican Jay in the Santa Rita Mountains, Arizona. Condor 51: 241–249.
Extensive notes on nest building, incubation, and growth of young.

Gullion, G. W.
1956. Evidence of double-brooding in Gambel Quail. Condor 58: 232–234.
In Nevada, near Davis Dam; compares Arizona conditions.

———
1960. The ecology of Gambel's Quail in Nevada and the arid southwest. Ecol. 41: 518–536.
Contains considerable data on Arizona habitats.

Haldeman, J. R. and A. B. Clark
1969. Walnut Canyon: An example of relationships between birds and plant communities. Plateau 41: 164–177.
The list contains first-seen and last-seen dates; includes species known to nest there.

Hamilton, T. H.
1962. Species relationships and adaptations for sympatry in the avian genus *Vireo*. Condor 64: 40–68.
Includes remarks on Marshall's study of the Towhees (Condor 62: 49–64, 1960).

Hanna, W. C.
1935. Whitney's Elf Owl. Ool. 52: 102–103.
Notes on sets of eggs in Arizona.

———
1961. Turkey Vulture nesting in Pima County, Arizona. Condor 63: 419.

Hardy, J. W.
1961. Studies in behavior and phylogeny of certain New World Jays (Garrulinae). Univ. Kansas Sci. Bull. 42: 13–149.
Includes important data on Mexican Jays in the Chiricahua Mountains.

Hargrave, L. L.
1933. The Western Gnatcatcher also moves its nest. Wilson Bull. 45: 30–31.
In northern Arizona.

———
1935. Red-tailed Hawk kills young Turkey. Condor 37: 83.
Near Flagstaff, Arizona.

———
1936. Three broods of Red-backed Juncos in one season. Condor 38: 57–59.
Near Flagstaff.

Hastings, J. R.
1959. Vegetation change and arroyo cutting in southeastern Arizona. Jour. Ariz. Acad. Sci. 1: 60–67.
An important paper on early conditions that affected animal life.

Heath, R. G.
1969. Nationwide residues of organochlorine pesticides in wings of Mallards and Black Ducks. Pesticides Monitoring Jour. 3: 115–123.
Reported from all states.

Hendrickson, J. R.
1949. Behavior of birds during a forest fire. Condor 51: 229–230.
In the Rincon Mountains, Arizona.

Henninger, W. F.
1899. Note on the Spotted Screech Owl *(Megascops trichopsis).* Osprey 4: 29.
In the Huachuca Mountains, Arizona.

Hensley, M. M.
1954. Ecological relations of the breeding bird population of the desert biome of Arizona. Ecol. Monog. 24: 185–207.
A study made in the Organ Pipe Cactus National Monument, Arizona.

———
1959. Notes on the nesting of selected species of birds of the Sonoran desert. Wilson Bull. 71: 86–92.
Accounts of 16 species in the Organ Pipe Cactus National Monument, Arizona.

Hickey, J. J., ed.
1969. Peregrine Falcon populations. Their biology and decline. Univ. Wisc. Press, Madison. 596 pp.
Comments on Arizona specimens.

[Hoffman, R.]
1924. [Describes voice of the Elf Owl near Tucson.] [In] Minutes of Cooper Club Meetings, Condor 26: 208.

Hoffman, W. J.
1876. Habits of western birds. Amer. Nat. 10: 238–239.
Crows and Ravens observed on the cliffs in the Grand Canyon.

Holterhoff, G., Jr.
1881. Verdin or Yellow-headed Titmouse. Ornith. and Ool. 6: 27.
Nesting notes from Arizona.

———
1883. Nest and eggs of Leconte's Thrasher *(Harporhynchus lecontei).* Bull. Nuttall Ornith. Club 8: 48–49.
At Flowing Wells, on the Southern Pacific Railroad, about 75 miles north of Fort Yuma.

Howard, O. W.
1899. Summer resident warblers of Arizona. Bull. Cooper Ornith. Club 1: 37–40; 63–65.
Field observations on 9 species of warblers.

———
1899. Some of the summer flycatchers of Arizona. Bull. Cooper Ornith. Club 1: 103–107.
Observations on the Sulphur-bellied, Olivaceous, and Buff-breasted Flycatchers.

———
1900. Nesting of the Mexican Wild Turkey in the Huachuca Mts., Ariz. *(Meleagris gallopavo).* Condor 2: 55–57.

———
1900. Nesting of the Rivoli Hummingbird in southern Arizona. Condor 2: 101–102.
In the Huachuca Mountains.

———
1902. Nesting of the Prairie Falcon. Condor 4: 57–59.
In the Huachuca Mountains, Arizona.

———
1904. The Coues Flycatcher as a guardian of the peace. Condor 6: 79–80.
Its breeding habits in the Huachuca Mountains, Arizona.

Howell, A. B.
1923. The influences of the southwestern deserts upon the avifauna of California. Auk 40: 584–592.
Discusses migration, changes in the avifauna from irrigation, and the desert as a barrier.

Hubbard, J. P.
1965. Bad days for the Black Hawk. Aud. Field Notes 19: 474.
Comments on the destruction of its cottonwood bosque habitat.

Huey, L. M.
1931. Skunks as prey for Owls. Wilson Bull. 43: 224.
Near Tucson, Arizona.

———
1932. Note on the food of an Arizona Spotted Owl. Condor 34: 100–101.
In the Chiricahua Mountains.

———
1935. A pair of Phainopeplas. Bird-Lore 37: 401–404.
Nesting notes from Castle Dome, Yuma County.

Huey, L. M.
1944. Nesting habits of the Hooded Oriole. Condor 46: 297.
In the Organ Pipe Cactus National Monument, Arizona.

Hungerford, C. R.
1960. Water requirements of Gambel's Quail. Trans. 25th N. A. Wildl. Conf. March 7, 8, 9, 1960: 231–240.
They can subsist well without free water by utilizing moist, succulent plant food.

———
1962. Adaptations shown in selection of food by Gambel Quail. Condor 64: 213–219.
In the Tucson region.

———
1964. Vitamin A and productivity in Gambel's Quail. Jour. Wildl. Mgmt. 28: 141–147.
The breeding rate changes with habitat quality; green plant food necessary.

Hunt, B. Mrs.
1959. "Insecticides and Birds" arouses Arizona reader. [In Letters in] Aud. Mag. 61: 194.
The University of Arizona received a grant of $2000 to study the effects of insecticides on birds in southern Arizona.

Irby, H. D. and L. H. Blankenship
1966. Breeding behavior of immature Mourning Doves. Jour. Wildl. Mgmt. 30: 598–604.
In the Tucson area.

Johnson, R. D.
1964. Red-tailed Hawk preys on black-tailed rattlesnake. Auk 81: 435.
In western Pima County, Arizona.

Judson, W. B.
1898. Nesting of Virginia's Warbler. Osprey 3: 54–55.
In the "mountains of southern Arizona."

Kalmbach, E. R.
1940. Economic status of the English Sparrow in the United States. U.S. Dept. Agric. Tech. Bull. 711: 1–66.
One stomach specimen listed from Arizona.

Keith, J. O.
1965. The Abert Squirrel and its dependence on ponderosa pine. Ecol. 46: 150–163.
Northern Arizona. Hawks prey on the squirrels, but the species is not identified.

Kennard, F. H.
1923. An owl's egg in the nest of a Western Red-tailed Hawk. Auk 40: 125.
North of the Santa Catalina Mountains, Arizona.

Kimball, H. H.
1925. Pygmy Owl killing a Quail. Condor 27: 209–210.
At the Portal Ranger Station, Chiricahua Mountains, Arizona.

———

1935. Elf Owls. Ool. 52: 95–96.

At Tucson and the Chiricahua Mountains; sets never contain more than three eggs.

Klopfer, P. H. and R. H. MacArthur

1960. Niche size and faunal diversity. Amer. Nat. 94: 293–300.

In Arizona 37.5 per cent of individuals are non-passerine. Hensley's 1954 figures are used in estimating territorial males per 100 acres.

Knipe, T.

1944. The status of the antelope herds of northern Arizona. Ariz. Game and Fish Comm. Pittman-Robertson Proj. Ariz. 9R: 1–40.

"Eagles take many fawns. . . ."

———

[195–]. The javelina in Arizona. Ariz. Game and Fish Dept. Wildl. Bull. 2: 1–96.

". . . they are not commonly accused of robbing quail nests, although it could be that some individuals acquire the habit and become habitual egg eaters."

Kraus, P.

1933. An outrage. Grand Canyon Nature Notes 8: 209–210.

A Pygmy Nuthatch kills a chipmunk that invaded its nest hole.

Kuhn, U. R.

1935. A food habit of the English Sparrow. Condor 37: 284–285.

At Nogales, ant eggs are eaten.

Kuykendall, J. R.

1963. Grapes are not for the birds. Prog. Agric. Ariz. 15 (6): 15.

Describes use of netting at the University of Arizona vineyard.

Ladd, S. B.

1891. Description of the nests and eggs of *Dendroica graciae* and *Contopus pertinax*. Auk 8: 314–315.

From Yavapai County, Arizona.

Lange, K. I.

1959. Soricidae of Arizona. Amer. Midl. Nat. 61: 96–108.

Remains found in pellets of Long-eared Owl, Barn Owl, and Horned Owl.

———

1959. Mammal remains from owl pellets in Arizona. Jour. Mamm. 40: 607.

Pellets from Long-eared Owl, Barn Owl, and Horned Owl from various localities.

Lanyon, W. E.

1963. Experiments on species discrimination in *Myiarchus* Flycatchers. Amer. Mus. Novit. 2126: 1–16.

At Portal, Arizona.

———

1963. Notes on a race of the Ash-throated Flycatcher, *Myiarchus cinerascens pertinax* of Baja California. Amer. Mus. Novit. 2129: 1–7.

Contains a sound spectrogram of *M. c. cinerascens* from Portal, Arizona.

Law, J. E.
1929. A discussion of faunal influences in southern Arizona. Condor 31: 216–220.
Disagrees with Swarth's 1929 "Faunal areas of southern Arizona." Proc. Calif. Acad. Sci. 4th ser. 18 (12): 267–383.

Lee, D. T. and E. Yensen
1969. Winter bird-population study: Riparian Woodland: Oak-Juniper Association. Aud. Field Notes 23: 538.
In Madera Canyon, Pima and Santa Cruz counties, Arizona.

——— and E. Yensen
1969. Winter bird-population study: Desert Scrub: Whitethorn Association. Aud. Field Notes 23: 543–544.
Near Continental, Arizona.

Leopold, A.
1918. Do Purple Martins inhabit bird boxes in the west? Condor 20: 93.
Nesting in pine snags at Lake Mary and Coleman Lake in Coconino County, Arizona.

Leopold, A. S. and R. A. McCabe
1957. Natural history of the Montezuma Quail in Mexico. Condor 59: 3–26.
Includes comparisons with Arizona birds.

Ligon, J. D.
1968. The biology of the Elf Owl, *Micrathene whitneyi.* Misc. Publ. Mus. Zool. Univ. Mich. 136: 1–70.
Field work chiefly in Cave Creek Canyon, Chiricahua Mountains, Arizona.

———
1968. Observations on Strickland's Woodpecker, *Dendrocopos stricklandi.* Condor 70: 83–84.
Includes observations on *D. arizonae* in Arizona.

———
1968. Starvation of spring migrants in the Chiricahua Mountains, Arizona. Condor 70: 387–388.
Freezing weather reduced the insect population.

———
1968. Sexual differences in foraging behavior in two species of *Dendrocopos* Woodpeckers. Auk 85: 203–215.
Arizona Woodpeckers were studied in Cave Creek Canyon, Cochise County, Arizona.

———
1969. Factors influencing breeding range expansion of the Azure Bluebird. Wilson Bull. 81: 104–105.
Scarcity of nest sites important.

———
1969. Some aspects of temperature relations in small owls. Auk 86: 458–472.
Three species of owls from the Chiricahua Mountains were used in the study.

Ligon, J. S.
1926. Habits of the Spotted Owl *(Syrnium occidentale)*. Auk 43: 421–429.
Summarizes published nesting records and discusses egg color.

Lowe, C. H., Jr.
1955. Gambel Quail and water supply on Tiburon Island, Sonora, Mexico. Condor 57: 244.
Mentions conditions in southern Arizona.

Lusk, R. D.
1899. Nesting of the Sulphur-bellied Flycatcher. Bull. Cooper Ornith. Club 1: 112–113.
Nests late in "southern Arizona."

——
1903. Wasted talent. Condor 5: 135.
Description of the nest of *Myiarchus l. olivascens*.

MacArthur, R. H.
1959. On the breeding distribution pattern of North American migrant birds. Auk 76: 318–325.
Includes an analysis of Hensley's data (Ecol. Monog. 24: 185–207, 1954) of breeding and migrant birds in the Organ Pipe Cactus National Monument.

——
1964. Environmental factors affecting bird species diversity. Amer. Nat. 98: 387–397.
The results were tested on the slopes of the Chiricahuas and at Tucson, but they did not always agree.

Marshall, J. T., Jr.
1960. Interrelations of Abert and Brown Towhees. Condor 62: 49–64.
Observations at Tucson, Arizona.

——
1962. Land use and native birds of Arizona. Jour. Ariz. Acad. Sci. 2: 75–77.
Advocates controlled burning of grass and forest litter to improve water capture and open the forest stands.

——
1963. Fire and birds in the mountains of southern Arizona. [In] Proc. 2nd Ann. Tall Timbers Fire Ecology Conf. March 14–15, 1963, pp. 134–142.
Discusses effects of fire on bird distribution in Arizona and northern Mexico.

——
1963. Land use and native birds of Arizona. Ariz. Cattlelog 19 (6): 14–15.
Advocates controlled burning to promote grass and water capture.

——
1963. Rainy season nesting in Arizona. [In] Proc. 13th Int. Ornith. Cong. Ithaca, 17–24 June, 1962: 2: 620–622.
Discusses nesting of Abert's Towhee.

——
1964. Voice in communication and relationships among Brown Towhees. Condor 66: 345–356.
A report based partly upon observations in the Tucson, Arizona, region.

Marshall, J. T., Jr.
1967. Parallel variation in North and Middle American Screech-owls. West. Found. Vert. Zool. Monog. 1: 1–72.
An exhaustive study, including Arizona species.

Martin, W. E.
1969. Organochlorine insecticide residues in Starlings. Pesticides Monitoring Jour. 3: 102–114.
Found in Arizona Starlings.

McAtee, W. L.
1911. Woodpeckers in relation to trees and wood products. U.S. Dept. Agric. Biol. Surv. Bull. 39: 1–99.
Lists trees damaged by Yellow-bellied Sapsucker in Arizona.

McCabe, T. T.
1936. Endemism and the American Northwest. Wilson Bull. 48: 289–302.
Most of Arizona is included in the region of greatest endemism.

McGregor, S. E., S. M. Alcorn, and G. Olin
1962. Pollination and pollinating agents of the saguaro. Ecol. 43: 259–267.
Includes birds that visit the saguaro.

McHenry, D. E.
1933. A broken-winged Robin. Grand Canyon Nature Notes 8: 178–181.
Rears a brood successfully at the Grand Canyon.

McKee, B. H.
1934. Raven vs. squirrel. Grand Canyon Nature Notes 8: 247.
The squirrel won possession of feeding tray.

———

1934. Some habits of two Ravens. Grand Canyon Nature Notes 9: 331–333.
At the Grand Canyon Village.

McKee, E. D.
1934. Pinyon nuts as bird feed. Grand Canyon Nature Notes 9: 272–273.

———

1934. [Notes on Chipping Sparrow, Goshawk, and White-throated Swift at the Grand Canyon.] Grand Canyon Nature Notes 9: 280–281.

McKelvey, M.
1965. Unusual bathing habits of the Turkey Vulture. Condor 67: 265.
Near Flagstaff, Arizona.

Merz, R. L.
1963. Jaw musculature of the Mourning and White-winged Doves. Univ. Kansas Publ. Mus. Nat. Hist. 12: 521–551.
The lengthened bill of the White-winged Dove is considered an adaptation for nectar-feeding at the saguaro cacti.

Miller, A. H.
1936. Tribulations of thorn-dwellers. Condor 38: 218–219.
Examples of accidents to Verdins, Palmer Thrashers, and Cactus Wrens in thorns and spines; near Tucson and Picacho.

———
1936. [Report of a month in Arizona.] [In] Minutes of Cooper Club Meetings, Condor 38: 254.
Mentions birds that feed on the fruit of the saguaro cactus, overgrazing in the Tumacacori Mountains, and studies of four species of Thrashers.

———
1937. The nuptial flight of the Texas Nighthawk. Condor 39: 42–43.
Near Tucson.

———
1937. Notes on the Saw-whet Owl. Condor 39: 130–131.
In the Sierra Ancha and Chiricahua mountains, Arizona; description of call notes.

———
1949. Some ecologic and morphologic considerations in the evolution of higher taxonomic categories. [In] Ornithologie als biologische Wissenschaft, 28 Beiträge als Festschrift zum 60 Geburtstag von Erwin Stresemann (22 November, 1949). Carl Winter, Heidelberg, Germany: 84–88.
Includes a discussion of the four species of Thrashers in Arizona and their modifications.

———
1956. Ecologic factors that accelerate formation of races and species of terrestrial vertebrates. Evolution 10: 262–277.
Includes distribution maps of *Melospiza melodia* and *M. lincolni.*

———
1960. Adaptation of breeding schedule to latitude. [In] Proc. 12th Int. Ornith. Cong. 2: 513–522.
In Arizona: *Cyrtonyx montezumae* and *Amphispiza bilineata* are discussed.

Miller, L.
1925. Food of the Harris Hawk. Condor 27: 71–72.
At Yuma.

———
1932. [Miller reports that Mearns Quail in Arizona hatched young in late August.] [In] Minutes of Cooper Club Meetings, Condor 34: 109.

———
1943. Notes on the Mearns Quail. Condor 45: 104–109.
Remarks on osteology, the breeding in Arizona, and natural enemies.

———
1952. Auditory recognition of predators. Condor 54: 89–92.
Imitated notes of owls were used to attract numbers of Arizona birds.

———
1962. High-noon songs. Condor 64: 75–76.
Whisper song of the Cassin Kingbird heard in the Pajarito Mountains, Arizona.

Miller, R. R. and H. E. Winn
1951. Observations on fish-eating by the Great-tailed Grackle in southeastern Arizona. Wilson Bull. 63: 207–208.

Monson, G.
1941. The effect of revegetation on the small bird population in Arizona. Jour. Wildl. Mgmt. 5: 395–397.
On the Navajo Indian Reservation.

———
1946. Roadrunner preys on Poor-will. Wilson Bull. 58: 185.
On the Cabeza Prieta Game Range, Yuma County, Arizona.

———
1951. Great Blue Heron killed by bobcat. Wilson Bull. 63: 334.
At Havasu Lake, Arizona.

———
1964. Ornithological aspects of Merriam's 1889 studies as viewed 75 years later. Plateau 37: 56–60.

Moore, M. D.
1928. The Cactus Wren. Bird-Lore 30: 106–107.
An account of its nesting in southern Arizona.

Moore, N. and J. Wahlstrom
1969. Winter bird-population study: Creosote-bush Desert. Aud. Field Notes 23: 544.
Near Tucson.

Moore, R. T.
1936. Description of a new race of *Carpodacus mexicanus*. Condor 38: 203–208.
Mentions habits of *C. m. frontalis* in Arizona.

Neff, J. A.
1940. Notes on nesting and other habits of the Western White-winged Dove in Arizona. Jour. Wildl. Mgmt. 4: 279–290.

———
1941. Arboreal nests of the Gambel Quail in Arizona. Condor 43: 117–118.
Nests in trees near Phoenix.

———
1943. Homing instinct in the Dwarf Cowbird in Arizona. Bird-Banding 14: 1–6.
Near Phoenix.

———
1944. Seeds of legumes eaten by birds. Condor 46: 207.
List of food items of White-winged Dove in Arizona.

———
1944. A protracted incubation period in the Mourning Dove. Condor 46: 243.
At Phoenix, Arizona.

———
1945. Foster parentage of a Mourning Dove in the wild. Condor 47: 39–40.
Mourning Dove broods White-winged Dove nestlings on the Gila River Indian Reservation.

———
1947. Habits, food, and economic status of the Band-tailed Pigeon. N. Amer. Fauna 58: 1–76.
Contains considerable data on Arizona birds.

Nice, M. M.
1923. A study of the nesting of Mourning Doves. Auk 40: 37–58.
Mentions two March nestings in Arizona.

N[orris], J. P.
1889. A series of eggs of Bendire's Thrasher. Ornith. and Ool. 14: 23–25.
Detailed descriptions of 29 nests and sets of eggs collected in Pima and Pinal counties, Arizona.

———
1889. Eggs of the Mexican Ground Dove. Ornith. and Ool. 14: 59–60.
Collected near Tucson.

———
1890. A series of eggs of Palmer's Thrasher. Ornith. and Ool. 15: 154–156.
Descriptions of 21 nests and sets of eggs from the vicinity of Tucson.

Nutting, W. L.
1966. Colonizing flights and associated activities of termites. I. The desert dampwood termite *Paraneotermes simplicicornis* (Kalotermitidae). Psyche 73: 131–149.
At Tucson, Nighthawks were present during termite twilight flights.

Oberholser, H. C.
1906. The North American Eagles and their economic relations. U.S. Dept. Agric. Biol. Surv. Bull. 27: 1–31.
Includes distributional maps.

O'Connor, J.
1939. Game in the desert. Derrydale Press, New York. 298 pp.
Some data on Turkey, Quail, and Dove habits and habitats.

———
1945. Hunting in the southwest. Alfred A. Knopf, New York. 279 pp. Originally published as Game in the Desert.
Information on habits and distribution of Turkey, Gambel's Quail, Mearns' Quail, White-winged Dove in Arizona.

Ohmart, R. D.
1969. Dual breeding ranges in Cassin Sparrow *(Aimophila cassinii)*. Abstract. P. 105 [in] C. C. Hoff and M. L. Riedesel, eds. Physiological systems in semiarid environments. Albuquerque, Univ. New Mexico Press.

Parker, H. G.
1886. Nest and eggs of the Plumbeous Gnatcatcher. Ornith. and Ool. 11: 54.
Breeding in Pinal County, Arizona.

———
1887. Notes on the eggs of the Thrushes and Thrashers. Ornith. and Ool. 12: 69–73.
Includes descriptions of the eggs of several species of Thrashers in Arizona.

Pase, C. P.
1969. Survival of *Quercus turbinella* and *Q. emoryi* seedlings in an Arizona chaparral community. S. W. Nat. 14: 149–155.
At Sierra Ancha Experimental Forest, Scrub Jays fed on acorns and cached them.

Pearson, G. A.
1913. Methods of combating seed-destroying animals. U.S. Dept. Agric. Fort Valley Exp. Sta. Review Forest Serv. Invest. 2: 82–85.
Junco dorsalis eats pine seeds and cuts seedlings; control is difficult.

Pearson, T. G.
1920. The Ground Dove. Bird-Lore 22: 126–129.
Contains notes on habits and a photograph of a nest of the Mexican Ground Dove at Tucson.

Peterson, R. T.
1948. Arizona Junco. Wilson Bull. 60: 5, with frontispiece of Arizona Junco in color.
Brief notes on habits and relationships.

Phillips, A. R.
1937. A nest of the Olive-sided Flycatcher. Condor 39: 92.
Near Flagstaff, Arizona.

———
1970. Avifauna in Mexico, pp. 69–74, [in] H. K. Buechner and J. H. Buechner, eds. The avifauna of northern Latin America. A symposium held at the Smithsonian Institution 13–15 April 1966. Smiths. Inst. Press, Wash. 119 pp.
A brief comment (p. 71) that Cowbird parasitism has caused local disappearances of some species in Arizona.

Poling, O. C.
1890. Nesting of the Arizona Jay. Ornith. and Ool. 15: 139.
In the Huachuca Mountains.

———
1890. On the nesting habits and eggs of the Vermilion Flycatcher. Ornith. and Ool. 15: 140.
Near Fort Huachuca.

———
1890. Notes on *Eugenes fulgens*. Auk 7: 402–403.
In the Huachuca Mountains.

Price, H. F.
1936. Comment. Ool. 53: 39–40.
Mentions a four egg set of Elf Owl from Tucson in reply to H. H. Kimball.

Price, W. W.
1888. Nesting of the Red-faced Warbler *(Cardellina rubrifrons)* in the Huachuca Mountains, southern Arizona. Auk 5: 385–386.

———
1888. Xantus's Becard *(Platypsaris albiventris)* in the Huachuca Mountains, southern Arizona. Auk 5: 425.

———

1895. The nest and eggs of the Olive Warbler *(Dendroica olivacea)*. Auk 12: 17–19.

Observations in the Huachuca, Chiricahua, Graham and White mountains.

Pugh, E. A.

1954. An unusual Goldfinch nest. Plateau 27: 22.

Near Flagstaff.

Radke, E. L. and E. R. Jones

1969. [Breeding bird census.]. Mesquite-palo verde-saguaro desert in Lower Sonoran Zone. Aud. Field Notes 23: 724.

Near Cave Creek, 25 miles north of Scottsdale, Arizona.

Rand, A. L.

1941. Courtship of the Roadrunner. Auk 58: 57–59.

Near Tucson.

———

1941. Arizona expedition. Nat. Hist. 48: 232–235.

In the Tucson area from January to June 1940, with brief notes on birds.

———

1941. Development and enemy recognition of the Curve-billed Thrasher *(Toxostoma curvirostre)*. Bull. Amer. Mus. Nat. Hist. 78: 213–242.

At Tucson.

———

1943. Some irrelevant behavior in birds. Auk 60: 168–171.

Examples from Tucson.

——— and R. M. Rand

1943. Breeding notes on the Phainopepla. Auk 60: 333–341.

Near Tucson.

Ransom, W. H.

1954. Some avian appetites. Part 2. News from the Bird-banders 29: 2–4.

Includes notes on food habits of birds at a Tucson, Arizona, feeding station.

Rasmussen, D. I.

1941. Biotic communities of Kaibab Plateau, Arizona. Ecol. Monog. 11: 229–275.

Lists 131 species of birds.

Ray, R. C.

1925. Discovery of a nest and eggs of the Blue-throated Hummingbird. Condor 27: 49–51.

In the Huachuca Mountains.

Rea, A. M.

1970. Winter territoriality in a Ruby-crowned Kinglet. Western Bird Bander 45: 4–7.

Mentions G. Austin's report of Kinglets being gregarious in the Santa Catalina Mountains, Arizona.

Redburn, R. A.

1932. Food for the Hawk. Grand Canyon Nature Notes 7: 40.

Sparrow Hawk captures Rock Squirrel.

Reynolds, H. G.
1951. Red-tailed Hawk captures cottontail rabbit. Condor 53: 151–152.
On the Santa Rita Experimental Range, south of Tucson.

———
1963. Western Goshawk takes Abert Squirrel in Arizona. Jour. Forestry 61: 839.

——— and R. R. Johnson
1964. Habitat relations of vertebrates of the Sierra Ancha Experimental Forest. U.S. Forest Serv. Res. Paper RM-4: 1–16.
Includes a list of 125 species of birds.

Richardson, F.
1938. Red Crossbills feeding on juniper galls. Condor 40: 257.
At the Grand Canyon, Arizona.

Ricklefs, R. E.
1965. Brood reduction in the Curve-billed Thrasher. Condor 67: 505–510.
At Tucson, Arizona.

———
1966. Behavior of young Cactus Wrens and Curve-billed Thrashers. Wilson Bull. 78: 47–56.
In the Tucson, Arizona, region.

———
1967. A case of classical conditioning in nestling Cactus Wrens. Condor 69: 528–529.

———
1968. The survival rate of juvenile Cactus Wrens. Condor 70: 388–389.

———
1969. An analysis of nesting mortality in birds. Smiths. Contr. Zool. 9: 1–48.
Includes Cactus Wrens in Arizona.

——— and F. R. Hainsworth
1967. The temporary establishment of dominance between two hand-raised juvenile Cactus Wrens *(Campylorhynchus brunneicapillus).* Condor 69: 528.

Rising, J. D.
1965. Notes on behavioral responses of the Blue-throated Hummingbird. Condor 67: 352–354.
In the Chiricahua Mountains, Arizona.

Root, R. B.
1967. The niche exploitation pattern of the Blue-gray Gnatcatcher. Ecol. Monog. 37: 317–350.
Winter studies made near Yuma and Tucson, Arizona.

———
1969. The behavior and reproductive success of the Blue-gray Gnatcatcher. Condor 71: 16–31.
At Yuma and Tucson.

Ross, A.

1967. Ecological aspects of the food habits of insectivorous bats. Proc. W. Found. Vert. Zool. 1: 205–263.

Page 255: Brief comments on Van Gelder and Goodpaster (Jour. Mamm. 33: 491, 1952) on competition in Arizona between Violet-green Swallows and bats.

———

1969. Ecological aspects of the food habits of insectivorous Screech Owls. Proc. W. Found. Vert. Zool. 1: 301–344.

Includes *Otus asio, O. trichopsis,* and *O. flammeolus* in Arizona.

Royall, W. C., Jr.

1962. Birds' deportment subject of study. Prog. Agric. Ariz. 14 (2): 3.

A general view of bird damage to crops.

———

1962. Starlings do damage to crops in Arizona. Prog. Agric. Ariz. 14 (3): 14–15.

Chiefly in the Phoenix area.

———

1966. Studies on bird depredation in Arizona. Pp. 96–97 [in] Proc. 5th annual meeting The Wildlife Soc. New Mexico-Arizona section, Showlow, Ariz., Feb. 4–5, 1966: 1–97.

Some damage by White-winged Doves reported in safflower fields.

———

1966. Breeding of the Starling in central Arizona. Condor 68: 196–205.

In the Salt River Valley.

Russell, H. N., Jr., and D. Amadon

1938. A note on highway mortality. Wilson Bull. 50: 205–206.

Includes northern Arizona species.

——— and A. M. Woodbury

1941. Nesting of the Gray Flycatcher. Auk 58: 28–37.

Nesting data from "northeastern Arizona."

Salt, G. W.

1952. The relation of metabolism to climate and distribution in three finches of the genus *Carpodacus*. Ecol. Monog. 22: 121–152.

Includes maps of breeding ranges of *C. cassinii* and *C. mexicanus.*

Scott, W. E. D.

1885. On the breeding habits of some Arizona birds. First paper. *Icterus parisorum.* Auk 2: 1–7.

In Pinal County.

———

1885. On the breeding habits of some Arizona birds. Second paper. *Icterus cucullatus.* Auk 2: 159–165.

In Pinal County.

———

1885. On the breeding habits of some Arizona birds. Third paper. *Phainopepla nitens.* Auk 2: 242–246.

In Pinal County.

Scott, W. E. D.
1885. On the breeding habits of some Arizona birds. Fourth paper. *Vireo vicinior*. Auk 2: 321–326.
In Pinal County.

———
1886. On the breeding habits of some Arizona birds. Fifth paper. *Aphelocoma sieberii arizonae*. Auk 3: 81–86.
Includes also *Peucaea ruficeps boucardi* and *Lophophanes wollweberi*.

Sheffler, W. J.
1937. [Reports late nesting of Band-tailed Pigeons in the White Mountains, Arizona.] [In] Minutes of Cooper Club Meetings, Condor 39: 95.

Shepardson, D. I.
1916. Some western birds—Cactus Wren. Ool. 33: 204–205.
Has two and three broods in Arizona; eggs were taken in August.

Short, L. L., Jr., and R. C. Banks
1965. Notes on the birds of northwestern Baja California. Trans. San Diego Soc. Nat. Hist. 14: 41–52.
Includes notes on the Starling nesting in Arizona.

Shufeldt, R. W.
1917. Interesting nests and eggs of some western birds. Ool. 34: 209–215.
Includes Coues' Flycatcher and Plumbeous Gnatcatcher from Arizona.

Smith, E. L.
1970. Cactus Wrens attack Ground Squirrel. Condor 72: 363–364.
In the Saguaro National Monument, east of Tucson, Arizona.

——— and K. M. Horn
1969. Winter bird-population study: Evergreen oak woodland. Aud. Field Notes 23: 537–538.
In Molino Basin picnic area in the Santa Catalina Mountains, 22 miles northeast of Tucson, Arizona.

Smith, L. M.
1941. Rock Wren nesting in a petrified log. Condor 43: 248.
At the Petrified Forest National Monument, Arizona.

Smith, P. W.
1900. Nesting of Stephen's Whippoorwill. Osprey 4: 89.
In the Huachuca Mountains, Arizona.

Smith, W. J.
1966. Communication and relationships in the genus *Tyrannus*. Nuttall Ornith. Club Publ. 6: 1–250.
Tyrannus verticalis, T. crassirostris, and *T. vociferans* were studied in southern Arizona.

———
1967. Displays of the Vermilion Flycatcher *(Pyrocephalus rubinus)*. Condor 69: 601–605.
Includes observations in southeastern and south-central Arizona.

Sperry, C. C.
1941. Food habits of the coyote. U.S. Dept. Int. Fish and Wildl. Serv. Wildl. Res. Bull. 4: 1–70.
Quail predation in Arizona discussed briefly.

Stafford, E. F.
1912. Notes on Palmer's Thrasher *(Toxostoma curvirostre palmeri).* Auk 29: 363–368.
Nesting habits at Tucson.

Stager, K. E.
1965. An exposed nocturnal roost of migrant Vaux Swifts. Condor 67: 81–82.
Two miles south of Davis Dam, Mohave County, Arizona.

Stannard, C.
1938. [Sparrow Hawk kills bird in trap.] News from the Bird-Banders 13: 45.
At Phoenix.

Steenbergh, W. F. and C. H. Lowe, Jr.
1969. Critical factors during the first years of life of the saguaro *(Cereus giganteus)* at Saguaro National Monument, Arizona. Ecol. 50: 825–834.
Page 829: seed removed by Curve-billed Thrasher. Suspected other birds: Gambel's Quail, Cactus Wren, and Brown Towhee.

Stevenson, J. O.
1951. September nesting of the Barn Swallow in Arizona. Wilson Bull. 63: 339–340.
At Springerville.

Stone, D. D.
1933. An Arizona apartment tree. Ool. 50: 110–111.
Seven species of birds in one mesquite tree at Casa Grande.

Stophlet, J. J.
1958. Hooded Oriole nesting under eaves of house. Auk 75: 221–222.
Near Tombstone, Arizona.

Strong, W. A.
1919. Curious eggs. Ool. 36: 180–181.
Includes Phainopepla and "Golden-cheeked Warbler . . . Ft. Small, Arizona," both with a Cowbird egg. Questionable.

———
1923. Large sets. Ool. 40: 64–70.
Several sets of eggs from Arizona are listed.

Sutton, G. M.
1953. Gray Hawk. Wilson Bull. 65: 5–7, with frontispiece sketch of hawk.
Brief mention of Arizona habitat.

Swank, W. G.
1952. Bird's eye view of a Mourning Dove factory. Ariz. Wildlife and Sportsman 23 (8): 32–39.
Excellent photographs of nesting.

Swarth, H. S.
1905. A correction. Condor 7: 144.
First set of *Setophaga picta* eggs collected by H. Brown in the Santa Rita Mountains, June 6, 1880.

———
1912. On the alleged egg-carrying habit of the Band-tailed Pigeon. Auk 29: 540–541.
Erroneous; neither are there any midwinter breeding records of this species in Arizona.

[Swarth, H. S.]
1928. [Eggs of Desert Quail in nest of Palmer Thrasher.] [In] Minutes of Cooper Club Meetings, Condor 30: 331.

Swinburne, J.
1890. The nest and eggs of *Regulus calendula.* Auk 7: 97–98.
In the White Mountains, Arizona.

Tainter, F. R.
1965. [Twenty-ninth breeding-bird census in] Cholla-palo verde-sahuaro foothill forest in Lower Sonoran Zone. Aud. Field Notes 19: 610–612.
Near Tucson, Arizona.

Taylor, W. K.
1966. Additional records of Black-tailed Gnatcatchers parasitized by the Dwarf Brown-headed Cowbird. Amer. Mid. Nat. 76: 242–243.
Near Mesa, Arizona.

Taylor, W. P.
1929. Order of awakening of some Arizona birds. Auk 46: 399.
Near Flagstaff.

——— and D. M. Gorsuch
1932. A test of some rodent and bird influences on western yellow pine reproduction at Fort Valley, Flagstaff, Arizona. Jour. Mamm. 13: 218–223.

Truett, J. C. and G. I. Day
1966. Winter food habits of coyotes and bobcats in Arizona. Pp. 83–87 [in] Proc. 5th annual meeting, The Wildlife Soc. New Mexico-Arizona section. Show Low, Arizona, Feb. 4–5, 1966: 1–97.
Gambel's Quail constituted .6 per cent of the food of the coyote, and 7.1 per cent of the food of the bobcat.

Udvardy, M. D. F.
1963. Bird faunas of North America. Proc. 13th Int. Ornith. Cong. Ithaca, 17–24, June, 1962. Vol. 2: 1147–1167.
Includes a brief account of the southwest region.

Underhill, R.
1940. The Papago Indians of Arizona and their relatives the Pima. U.S. Office Indian Affairs, Education Div. 1–68 pp.
Quail and Doves were eaten.

Van Gelder, R. G. and W. W. Goodpaster
1952. Bats and birds competing for food. Jour. Mamm. 33: 491.
Violet-green Swallows and Pipistrelles near Fort Grant, Arizona.

van Rossem, A. J.
1944. [Report of Kingfisher in Arizona catching a lizard.]. [In] Minutes of Cooper Club Meetings, Condor 46: 92.

Viers, C. E. and L. Sileo
1969. Winter bird-population study: Cholla Cactus-Creosote-bush Desert. Aud. Field Notes 23: 544–545.
Near Tucson.

Voigt, R. L.
1965. A bird-tolerant hybrid grain sorghum for Arizona. Prog. Agric. in Ariz. 17: 21–22.
Longer glumes and a greater percentage of tannic acid are apparently distasteful to English Sparrow, Doves, and Blackbirds.

———
1966. Crop protection—bird-tolerant Sorghum-crop pests. Crops and Soils 19: 25.
Bird depredations were extreme near Yuma; the species are not listed.

———
1966. Bird-tolerant sorghums boost take-home yields. Prog. Agric. in Ariz. 18: 30–32.
Losses from Sparrows and Doves decreased in plantings of the new bird-tolerant sorghums.

Vorhies, C. T.
1928. Do southwestern Quail require water? Amer. Nat. 62: 446–452.
Evidently they do not.

———
1928. Band-tailed Pigeon nesting in Arizona in September. Condor 30: 253.
In the Santa Catalina Mountains.

Vor[h]ies, C. T.
1929. Do southwestern Quail require water? Ariz. Wild Life Mag. 2: 17–18.
A summary of his paper in Amer. Nat. 62: 446–452, 1928.

Vorhies, C. T.
1934. The White-necked Raven, a change of status? Condor 36: 118–119.
The Ravens disappeared when the garbage dumps were removed.

———
1937. Inter-relationships of range animals. Trans. 2nd N. A. Wildl. Conf. March 1, 2, 3, 4, 1937: 288–294.
Discusses food of Roadrunner and Marsh Hawk in Arizona.

——— and W. P. Taylor
1933. The life histories and ecology of Jack Rabbits, *Lepus alleni* and *Lepus californicus* ssp., in relation to grazing in Arizona. Univ. Ariz. Agric. Exp. Sta. Tech. Bull. 49: 471–587.
Golden Eagles preyed on the antelope jack rabbits.

——— and W. P. Taylor
1940. Life history and ecology of the white-throated wood rat, *Neotoma albigula albigula* Hartley, in relation to grazing in Arizona. Univ. Ariz. Agric. Exp. Sta. Tech. Bull. 86: 455–529.
No evidence that it destroys bird nests.

Waesche, H. H.
1932. A baffling piece of plate glass. Grand Canyon Nature Notes 7: 55–57.
Birds killed by flying into a window.

Walker, L. W.
1943. Nocturnal observations of Elf Owls. Condor 45: 165–167.
At a nest in the Kofa Mountain Game Refuge, Yuma County, Arizona.

———
1949. Talons in the night. Aud. Mag. 51: 226–237.
Includes notes on the nesting of the Elf Owl in Arizona.

Wallmo, O. C.
1954. Nesting of Mearns Quail in southeastern Arizona. Condor 56: 125–128.
Includes photographs of nests in the Huachuca Mountains.

Watson, W. A.
1936. An unusual nesting habit. Grand Canyon Nat. Hist. Assoc. Bull. 4: 5–6.
A Chestnut-backed Bluebird and a Pygmy Nuthatch occupy the same nest cavity!

Westcott, P. W.
1964. Unusual feeding behavior of a Goshawk. Condor 66: 163.
Goshawk captures Abert squirrel in the Santa Catalina Mountains, Arizona.

———
1969. Relationships among three species of Jays wintering in southeastern Arizona. Condor 71: 353–359.
They kept apart, and ate the same food.

Wetmore, A.
1920. Observations on the habits of the White-winged Dove. Condor 22: 140–146.
Near Arlington, Arizona.

Wheeler, R. S.
1887. Rufus-vented or Crissal Thrasher; Bendire's Thrasher; and Canon Towhee. Ool. 4: 76.
Nesting notes from Pima Agency, Arizona.

White, H. G.
1890. Geographical variation of eggs. Ornith. and Ool. 15: 1–4.
Includes descriptions of several sets of eggs of Arizona species.

Whitfield, C. J.
1934. A Screech Owl captured by a snake. Condor 36: 84.
At the Parker Creek Experiment Station, Tonto National Forest, Arizona.

Wible, M.
1967. Wing and tail flashing of Painted Redstart. Wilson Bull. 79: 246.
Near Payson, Arizona.

Wilbur, ?
1894. [Least Vireo nests late in Arizona.] [In] Meeting of Cooper Ornith. Club, Nidiologist 1: 143.
As late as June 19, 1893.

Willard, F. C.
1898. Freezing to death (?). Osprey 2: 119.
A Brewer's Blackbird falls dead and is presumed to be frozen.

———
1898. Quails going to roost. Osprey 2: 134.
Scaled Quail near Tombstone.

———
1899. Notes on *Eugenes fulgens*. Osprey 3: 65–66.
Nesting in the Huachuca Mountains; photographs of nests.

———
1908. An Arizona nest census. Condor 10: 44–45.
Brief accounts of the nesting of 14 species of birds at Tombstone.

———
1908. Huachuca notes. Condor 10: 206–207.
Elevations of nests of some species of birds differ on the east and west sides of the range.

———
1908. Three Vireos: nesting notes from the Huachuca Mountains. Condor 10: 230–234.
Lanivireo solitarius plumbeus, V. huttoni stephensi, Vireosylva gilva swainsoni.

———
1909. Behavior of a young Rivoli Hummingbird. Condor 11: 102–103.
In the Huachuca Mountains.

———
1909. Nesting of the Arizona Junco. Condor 11: 129–131.
In the Huachuca Mountains.

———
1909. The Flammulated Screech Owl. Condor 11: 199–202.
Breeding in the Huachuca Mountains.

———
1910. Nesting of the Western Evening Grosbeak *(Hesperiphona vespertina montana).* Condor 12: 60–62.
In the Santa Catalina and Huachuca mountains.

———
1910. The Olive Warbler *(Dendroica olivacea)* in southern Arizona. Condor 12: 104–107.
Observations on nesting.

———
1911. The Blue-throated Hummingbird. Condor 13: 46–49.
Breeding in the Huachuca Mountains.

———
1912. Breeding of the Scott Sparrow. Condor 14: 195–196.
In the Huachuca Mountains.

———
1912. Nesting of the Rocky Mountain Nuthatch. Condor 14: 213–215.
In the Huachuca Mountains.

Willard, F. C.

1913. Some late nesting notes from the Huachuca Mountains, Arizona. Condor 15: 41.

Toxostoma curvirostre palmeri, Cyanolaemus clemenciae, and *Columba fasciata.*

———

1913. Late nesting of certain birds in Arizona. Condor 15: 227.

Observations on 7 species in the Huachuca Mountains.

———

1913. Sharp-shinned Hawk nesting in Arizona. Condor 15: 229.

In the Huachuca Mountains.

———

1915. A curious set of Gambel Quail eggs. Condor 17: 97.

Near Tucson.

———

1916. The Golden Eagle in Cochise County, Arizona. Ool. 33: 3–8.

Notes on nesting, with photographs.

———

1916. Nesting of the Band-tailed Pigeon in southern Arizona. Condor 18: 110–112.

In the Huachuca Mountains.

———

1918. Evidence that many birds remain mated for life. Condor 20: 167–170.

Gives many Arizona examples.

———

1923. Some unusual nesting sites of several Arizona birds. Condor 25: 121–125.

———

1923. The Mexican Cliff Swallow in Cochise County, Arizona. Condor 25: 138–139.

Nesting at Fort Huachuca.

———

1923. The Buff-breasted Flycatcher in the Huachucas. Condor 25: 189–194.

Nest and eggs collected in May 1907.

Willis, E. O.

1963. Is the Zone-tailed Hawk a mimic of the Turkey Vulture? Condor 65: 313–317.

Includes observations in southern Arizona.

[Wood, (?)]

1913. [Report of bird mortality in the vicinity of Prescott, Arizona.] [In] Minutes of Cooper Club Meetings, Condor 15: 236.

Woodbury, A. M., et al.

1959. Ecological studies of the flora and fauna in Glen Canyon. Univ. Utah Anthro. Papers, no. 40, Glen Canyon Ser. 7: 1–226.

Birds, pp. 107–133, by W. H. Behle and H. G. Higgins.

Woods, R. S.
1929. The Arizona Junco. Amer. Forests and Forest Life 35: 337.
Brief notes on nesting with photos of nest in the Santa Catalina Mountains.

Wright, G. M.
1932. A bat-eating Sparrow Hawk. Condor 34: 43.
In the Grand Canyon, Arizona.

6. Diseases and Parasites

Only papers dealing specifically with Arizona birds are listed in this small section.

Blankenship, L. H., R. E. Reed, and H. D. Irby
1966. Pox in Mourning Doves and Gambel's Quail in southern Arizona. Jour. Wildl. Mgmt. 30: 253–257.
In the Tucson area.

Campbell, H. and L. Lee
1953. Studies on Quail malaria in New Mexico and notes on other aspects of Quail populations. New Mexico Dept. Game and Fish, Santa Fe, New Mexico: 1–79.
A louse fly, *Microlynchia pusilla,* is reported from Scaled Quail in Arizona.

Emerson, K. C.
1948. Two new species of Mallophaga. Jour. Kansas Entomol. Soc. 21: 137–140.
Colinicola mearnsi from Mearns' Quail at Nogales, Arizona.

———
1949. Three new species of Mallophaga. Jour. Kansas Entomol. Soc. 22: 75–78.
Oxylipeurus montezumae from Mearns' Quail from Apache Indian Reservation; *Lagopoecus gambelii* from Gambel's Quail from Tucson, Arizona.

———
1950. The genus *Lagopoecus* (Philopteridae: Mallophaga) in North America. Jour. Kansas Entomol. Soc. 23: 97–101.
Lists *L. gambelii* from Gambel's Quail in Arizona.

———
1950. New species of *Goniodes*. Jour. Kansas Entomol. Soc. 23: 120–126.
G. submamillatus from Mearns' Quail in Arizona.

———
1951. A list of Mallophaga from gallinaceous birds of North America. Jour. Wildl. Mgmt. 15: 193–195.
Parasites from Quail and Turkey from Arizona are listed.

———
1953. New North American Mallophaga. Jour. Kansas Entomol. Soc. 26: 132–136.
Five species described from Arizona birds.

———
1955. Notes on two species of *Goniodes*. Jour. Kansas Entomol. Soc. 28: 146.
A correction: *G. submamillatus* came from Gambel's Quail (see Emerson, 1950, Jour. Kansas Entomol. Soc. 23: 120–126).

——— and J. C. Johnson, Jr.
1961. The genus *Penenirmus* (Mallophaga) found on North American Woodpeckers. Jour. Kansas Entomol. Soc. 34: 34–43.
Two species listed from Arizona birds.

——— and H. D. Pratt
1956. The Menoponidae (Mallophaga) found on North American Swifts. Jour. Kansas Entomol. Soc. 29: 21–28.
Two species from Arizona birds.

Erling, H. G.
1956. Disease-parasite inspection of Arizona wildlife. Ariz. Game and Fish Dept. Completion Rep. proj. no. W-53-R-6, work plan 8, job no. 1: 1–5.

Hungerford, C. R.
1955. A preliminary evaluation of Quail malaria in southern Arizona in relation to habitat and Quail mortality. Trans. 20th N. A. Wildlife Conf. March 14, 15, and 16, 1955: 209–219.

Kalmbach, E. R. and M. F. Gunderson
1934. Western duck sickness: a form of botulism. U.S. Dept. Agric. Tech. Bull. 411: 1–81.
Lists Picacho Reservoir, Pinal County, Arizona, as one of the localities where disease was present in 1928.

Smith, A. P.
1908. Brain parasite in the White-necked Raven. Condor 10: 92.
Parasite not identified.

Toepfer, E. W., Jr.
1964. *Colpoda steinii* in oral swabbings from Mourning Doves (*Zenaidura macroura* L.). Jour. Parasitol. 50: 703.
Some from birds from Pima County, Arizona.

———, L. N. Locke, and L. H. Blankenship
1966. The occurrence of *Trichomonas gallinae* in White-winged Doves in Arizona. Bull. Wildl. Disease Assoc. 2 (1): 13.
Near Tucson: first record in Arizona.

Wood, S. F. and C. H. Herman
1943. The occurrence of blood parasites in birds from southwestern United States. Jour. Parasitol. 29: 187–196.
Six species of Arizona birds are listed.

7. Game Management and Conservation

Most of the papers in this section have been published by the Arizona Game and Fish Department. Their work is rather extensive.

Allen, D. L., ed.
1956. Pheasants in North America. Stackpole Co. and Wildl. Mgmt. Inst. 490 pp. Chap. IV, pp. 159–203, Pheasants in the arid southwest, is by L. E. Yeager, J. B. Low, and H. J. Figge.
Not a success in Arizona.

Allen, F. H.
1941. Conservation notes. Auk 58: 288.
Comments on the "critical" situation of the White-winged Dove in Arizona.

Anonymous.
1928. [Nine Wild Turkeys from Arizona presented to the California Game Farm by George O'Connor.] Calif. Fish and Game 14: 165.
No locality given.

Anonymous.
1932. Along the trail. Ariz. Wild Life 4 (4): 21.
Three "Pheasants" released 11 miles north of Globe along Pinal Creek nested successfully.

Anonymous.
1935. Many Roadrunners slain by Yuma boys in contest. Ariz. Producer 14 (7): 21.
A total of 987 killed during a period of two months: 10 cents bounty paid.

[Anonymous.-ed.]
1938. Masked Bob White released in forest. Ariz. Wildlife Mag. 7 (8): 16.
In Coronado National Forest grass area, 66 released.

Anonymous.
1960. [Game management.] Small game. Wildlife Views 7 (1): 28–30.

Arizona Game and Fish Commission [Later author, Arizona Game and Fish Department.]
1949–1970. Information supporting recommendations on [year] fall hunting season. This annual publication, averaging about 50 pages, also sometimes titled "Data supporting recommendations. . . ," in 1970 was entitled "Data Summary. Arizona Game and Fish Department."

Arizona Game and Fish Department
1964. The reasons behind the 1964–65 hunting regulations. Wildlife Views 11 (3): 4–7.
Data on abundance of Chukar, Turkey, and Blue Grouse.

Arnold, L. W.
1942. The Gambel's Quail of the Yuma district. Ariz. Wildlife and Sportsman 4 (5): 4, 12.

Arrington, O. N. (Pop)
1950. 1949 Turkey. Arizona Wildlife and Sportsman 11 (5): 7, 12.
A summary of populations and harvest.

Barnes, [T. F.]
1967. Region V. [In annual report, 7-1-65 to 6-30-66.] Wildlife Views 14 (1): 43–45.
Gambel's Quail are "probably more numerous today than at any time in modern history." Phoenix area.

Berlinski, D.
1963. Arizona's waterfowl management areas. Wildlife Views 10 (6): 4–7.

Bishop, R. [A.]
1964. Population figures of Arizona's Mearns' Quail. Proc. 3rd Ann. Meeting Wildl. Soc. New Mex.-Ariz. Sec. Feb. 7, 8, 1964: 58–61.
In the Santa Rita Mountains and Canelo Hills.

——— and C. R. Hungerford
1965. Seasonal food selection of Arizona Mearns' Quail. Jour. Wildl. Mgmt. 29: 813–819.
Bulbs and tubers *(Oxalis* and *Cyperus)* form a high percentage of its food.

Blankenship, L. H., R. E. Tomlinson, and R. C. Kufeld
1967. Arizona dove wing survey, 1964. Bur. Sport Fisheries and Wildl. Spec. Sci. Rep. Wildl. 116: 1–34.
The survey was considered a success.

Boeker, E. L. and V. E. Scott
1969. Roost tree characteristics for Merriam's Turkey. Jour. Wildl. Mgmt. 33: 121–124.
Studies in the White Mountain Apache Indian Reservation, Arizona.

Bohl, W. H.
1968. Results of foreign game introduction. [In] Trans. 33rd N. Amer. Wildl. and Nat. Resources Conf. March 11–13, 1968: 389–398.
Brief mention of introduction of Afghan White-winged Pheasant in Arizona.

Bumstead, R.
1956. Investigation of Kaibab north. Ariz. Game and Fish Dept. Completion Rep. proj. no. W-53-R-6, work plan 5, job no. 3: 1–7.

Bureau of Sport Fisheries and Wildlife
1966. Wildlife research: problems, programs, progress, 1965. U.S. Dept. Int. Fish and Wildl. Serv. Bur. Sport Fisheries and Wildl. Resource Publ. 23: 1–102.
For Arizona: chiefly a summary of new publications relating to wildlife.

Cady, L. M.
1911. The conservation of bird life. Birds of Arizona. Ariz. The New State Mag. 1(5): 18–19.
A plea with some misinformation and misidentifications of Arizona species of birds.

Cahalane, V. H., et al.
1940. Report of the committee on bird protection, 1939. Auk 57: 279–291.
Mentions that the Masked Bob-white has been reintroduced into Arizona.

———
1942. Report of the committee on bird protection, 1941. Auk 59: 286–300.
Discusses the decline in the White-winged Dove in Arizona.

———
1943. Report of the committee on bird protection, 1942. Auk 60: 152–162.
Early migration of Arizona White-winged Dove results in negligible kill.

Carr, J. N.
1960. Mourning Dove and White-winged Dove nest surveys during the summer of 1960. Ariz. Game and Fish Dept. Suppl. Rep. proj. W-53-R-11, work plan 3, job no. 2, objective 4: 1–6.

Clarke, C. H. D., et al.
1965. Report of the committee on bird protection, 1964. Auk 82: 477–491.
Mentions briefly the attempts to reestablish the Masked Bobwhite in Arizona.

Committee on Rare and Endangered Wildlife Species
1966. Rare and endangered fish and wildlife of the United States. Bur. Sport Fisheries and Wildl. Resource Publ. 34: 1–180.
Lists a number of Arizona birds, some endangered, some periferal.

———
1968. Rare and endangered fish and wildlife of the United States. Bur. Sport Fisheries and Wildl. Rev. ed. 274 pp.

Cosper, P. M.
1949. Merriam's Turkey population trend survey technique. Ariz. Game and Fish Comm. Spec. Rep. proj. 25-R-3, job no. 1: 1–3.

Cottam, C. and J. B. Trefethen, eds.
1968. Whitewings—The life history, status, and management of the White-winged Dove. Van Nostrand and Co. Princeton. 348 pp.
A comprehensive account.

———, et al.
1942. Report of the committee on bird protection, 1941. Auk 59: 286–300.
They recommended a closed season on White-winged Doves in Arizona in order to build up the population.

Davis, W. C.
1962. Values of hunting and fishing in Arizona, 1960. Univ. Ariz. Bur. of Bus. and Pub. Res. Spec. Studies 21: 1–66.
Includes summary of bird harvest in 1960.

Day, G. I.
1955. Investigation of Tumacacori Mountains. Ariz. Game and Fish Dept. Completion Rep. proj. W-53-R-5, work plan 5, job no. 14: 1–6.
Includes Quail and Doves.

———
1955. Investigation of Whetstone area. Ariz. Game and Fish Dept. Completion Rep. proj. W-53-R-6, work plan 5, job no. 18: 1–9.

———
1956. Investigation of Canelo-Patagonia Mountains. Ariz. Game and Fish Dept. Completion Rep. proj. W-53-R-6, work plan 5, job no. 19: 1–8.

Diem, K. L.
1952. Investigation of Kaibab north. Ariz. Game and Fish Comm. Completion Rep. proj. 53-R-2, work plan 5, job no. 5: 1–8.
Reports increase of Turkeys.

———
1954. Investigation of Kaibab north. Ariz. Game and Fish Comm. Completion Rep. proj. W-53-R-4, work plan 5, job no. 3: 1–21.

Eicher, G. J., Jr.
1945. Picacho Lake survey, January to July, 1944. Ariz. Game and Fish Comm. Fed. Aid Div. Ariz. Wildl. and Sportsman 6 (8): 12–14.

———
1945. Midsummer Quail survey, July 5 to 31, 1944. Ariz. Game and Fish Comm. Fed. Aid Div. Ariz. Wildl. and Sportsman 6 (5): 12–13.

Erickson, R. C.
1968. A federal research program for endangered wildlife. [In] Trans. 33rd N. Amer. Wildl. and Nat. Resources Conf. March 11–13, 1968: 418–433.
Includes a review of progress in Masked Bobwhite restocking in Arizona.

———, et al.
1968. Report of committee on conservation, 1968. Auk 85: 669–677.
Masked Bobwhite will be released in the Altar Valley in Arizona.

Evans, T. R.
1964. Report of migratory bird committee. [In] 54th Conv. Int. Assoc. Game, Fish and Conservation Comms. pp. 95–100.
Mourning Dove, p. 98: wing collection for dove kill sampling in Arizona.

———
1966. Annual report of the Migratory Birds Committee. [In] 56th Conv. Int. Assoc. Game, Fish and Conservation Comms. pp. 34–43.
White-winged Dove, p. 39: breeding population estimated at 1,250,000 in Arizona.

Evenden, F. G., Jr.
1952. Waterfowl sex ratios observed in the western United States. Jour. Wildl. Mgmt. 16: 391–393.
Eight species from Arizona are listed.

Ferguson, J. W.
1951. Investigation of the Three Bar Game Management Area. Ariz. Game and Fish Comm. Completion Rep. proj. 53-R, work plan 5, job no. 6: 1–14.

Finley, W. L.
1910. Report of William L. Finley, in Annual report of the National Association of Audubon Societies for 1910. Bird-Lore 12: 262, 275–277.
Discusses slaughter of Doves in the Tucson area.

Fleming, W. B.
1952. Waterfowl concentration areas. Ariz. Game and Fish Comm. Completion Rep. proj. 53-R-2, work plan 3, job no. 10: 1–8.

———
1953. An inventory of waterfowl food plants in Arizona. Ariz. Game and Fish Comm. Completion Rep. proj. W-70-R-1, work plan 1, job no. 2: 1–8.

———
1953. Midwinter statewide waterfowl survey development. Ariz. Game and Fish Comm. Completion Rep. proj. W-70-R-1, work plan 1, job no. 3: 1–7.

———
1953. Waterfowl hunt data collection. Ariz. Game and Fish Comm. Completion Rep. proj. W-70-R, work plan 2, job no. 1: 1–7.

———
1953. Statewide waterfowl management research. Waterfowl trapping and banding. Ariz. Game and Fish Comm. Completion Rep. proj. W-70-R-1, work plan 4, job no. 1: 1–25.

———
1954. Inventory of existing waterfowl habitat and its utilization in Arizona. Ariz. Game and Fish Comm. Completion Rep. proj. W-70-R-1, work plan 1, job no. 1: 1–19.

———
1954. Waterfowl food habits study. Ariz. Game and Fish Comm. Completion Rep. proj. W-70-R-1, work plan 3, job no. 1: 1–5.

———
1954. Evaluation of developed areas. Ariz. Game and Fish Comm. Completion Rep. proj. W-70-R-1, work plan 5, job no. 1: 1–6.
For waterfowl.

———
1954. Waterfowl management recommendations. Ariz. Game and Fish Comm. Completion Rep. proj. W-70-R-1, work plan 6, job no. 1: 1–5.

———
1954. Inventory of existing waterfowl habitat and its utilization in Arizona. Ariz. Game and Fish Dept. Completion Rep. proj. W-70-R-2, work plan 1, job no. 1: 1–19, with maps.

———

1954. An inventory of waterfowl food plants in Arizona. Ariz. Game and Fish Dept. Completion Rep. proj. W-70-R-2, work plan 1, job no. 2: 1–4.

———

1954. Midwinter statewide waterfowl survey development. Ariz. Game and Fish Comm. Completion Rep. proj. W-70-R-2, work plan 1, job no. 3: 1–10.

———

1954. Waterfowl hunt data collection. Ariz. Game and Fish Comm. Completion Rep. proj. W-70-R-2, work plan 2, job no. 1: 1–9.

———

1954. Waterfowl trapping and banding. Ariz. Game and Fish Comm. Completion Rep. proj. W-70-R-2, work plan 4, job no. 1: 1–2.

———

1954. Evaluation of developed areas. Ariz. Game and Fish Comm. Completion Rep. proj. W-70-R-2, work plan 5, job no. 1: 1–10.

For waterfowl.

———

1954. Waterfowl management recommendations. Ariz. Game and Fish Dept. Completion Rep. proj. W-70-R-2, work plan 6, job no. 1: 1–7.

———

1955. Survey of waterfowl habitat and habitat utilization in Arizona. Ariz. Game and Fish Dept. Completion Rep. proj. W-70-R-3, work plan 1, job no. 1: 1–17.

———

1955. An inventory of waterfowl food plants in Arizona. Ariz. Game and Fish Dept. Completion Rep. proj. W-70-R-3, work plan 1, job no. 2: 1–11.

———

1955. Midwinter statewide waterfowl survey development. Ariz. Game and Fish Dept. Completion Rep. proj. W-70-R-3, work plan 1, job no. 3: 1–5.

———

1955. Waterfowl hunt data collection. Ariz. Game and Fish Dept. Completion Rep. proj. W-70-R-3, work plan 2, job no. 1: 1–8.

———

1955. Waterfowl trapping and banding. Ariz. Game and Fish Dept. Completion Rep. proj. W-70-R-3, work plan 4, job no. 1: 1–3.

———

1955. Evaluation of developed areas. Ariz. Game and Fish Dept. Completion Rep. proj. W-70-R-3, work plan 5, job no. 1: 1–12.

———

1955. Waterfowl management recommendations. Ariz. Game and Fish Dept. Completion Rep. proj. W-70-R-3, work plan 6, job no. 1: 1–11.

Contains considerable migration data.

Fleming, W. B.
1956. Midwinter statewide waterfowl survey development. Ariz. Game and Fish Dept. Completion Rep. proj. W-70-R-4, work plan 1, job no. 3: 1–4.

———
1956. Waterfowl hunt data collection. Ariz. Game and Fish Dept. Completion Rep. proj. W-70-R-4, work plan 2, job no. 1: 1–5.

———
1956. Evaluation of developed areas. Ariz. Game and Fish Dept. Completion Rep. proj. W-70-R-4, work plan 5, job no. 1: 1–14.

———
1956. Waterfowl management recommendations. Ariz. Game and Fish Dept. Completion Rep. proj. W-70-R-4, work plan 6, job no. 1: 1–4.

———
1957. Waterfowl hunt data collection. Ariz. Game and Fish Dept. Completion Rep. proj. W-70-R-5, work plan 2, job no. 1: 1–8.

———
1957. Waterfowl trapping and banding. Ariz. Game and Fish Dept. Completion Rep. proj. W-70-R-5, work plan 4, job no. 1: 1–7.

———
1957. Evaluation of developed areas. Ariz. Game and Fish Dept. Completion Rep. proj. W-70-R-5, work plan 5, job no. 1: 1–10.

———
1957. Waterfowl management recommendations. Ariz. Game and Fish Dept. Completion Rep. proj. W-70-R-5, work plan 6, job no. 1: 1–4.

———
1958. Midwinter statewide waterfowl survey development. Ariz. Game and Fish Dept. Completion Rep. proj. W-70-R-6, work plan 1, job no. 3: 1–3.

———
1958. Waterfowl hunt data collection. Ariz. Game and Fish Dept. Completion Rep. proj. W-70-R-6, work plan 2, job no. 1: 1–4.

———
1958. Waterfowl trapping and banding. Ariz. Game and Fish Dept. Completion Rep. proj. W-70-R-6, work plan 4, job no. 1: 1–8.

———
1959. Evaluation of developed areas. Ariz. Game and Fish Dept. Completion Rep. proj. W-70-R-6, work plan 5, job no. 1: 1–4.

———
1959. Migratory waterfowl in Arizona. Ariz. Game and Fish Dept. Wildlife Bull. 5: 1–74.
Data on habitats, food, nesting, migration, and hunting.

———
1959. Waterfowl. [In] Annual Report, July 1, 1957 to June 30, 1958, of Ariz. Game and Fish Dept. Wildlife News 6 (1): 22–24.
A summary of trapping and banding, including the January population survey.

———
1960. Waterfowl. Wildlife Views 7 (1): 33–34.
A summary of Game Department work.

———
1961. Waterfowl. [In] Annual Report, July 1, 1959 to June 30, 1960, of Arizona Game and Fish Dept. Wildlife Views 8 (1): 4.
Includes data on populations.

———
1962. Waterfowl. [In] Annual Report, July 1, 1960 to June 30, 1961, of Arizona Game and Fish Dept. Wildlife Views 9 (1): 6–8.
Includes data on populations and banding.

———
1963. Waterfowl. [In] Annual Report, July 1, 1961 to June 30, 1962, of Arizona Game and Fish Dept. Wildlife Views 10 (1): 12–13.
Geese have increased in Arizona.

———
1964. Waterfowl. [In] Annual Report, July 1, 1962 to June 30, 1963, of Arizona Game and Fish Dept. Wildlife Views 11 (1): 13.
Includes data on populations.

———
1965. Waterfowl. [In] Annual Report, July 1, 1963 to June 30, 1964, of Arizona Game and Fish Dept. Wildlife Views 12 (1): 13.
The wintering population of Canada Geese showed an increase.

———
1966. Waterfowl. [In] Annual Report, July 1, 1964 to June 30, 1965, of Arizona Game and Fish Dept. Wildlife Views 13 (1): 10.
Includes brief notes on development units for waterfowl near Painted Rock Dam and near Winslow.

———
1967. Waterfowl. [In] Annual Report, 7-1-65 to 6-30-66. Wildlife Views 14 (1): 11–12.
Canada Geese are introduced into the White Mountain area, Arizona.

———
1968. Waterfowl in Arizona. Wildlife Views 15 (5): 26–27.
Their migration and distribution are discussed briefly.

———
1968. Waterfowl. [In] Annual Report, 7-1-66 to 6-30-67, of Arizona Game and Fish Dept. Wildlife Views 15 (1): 10–11.

———
1969. Waterfowl. [In] Annual Report, 7-1-67 to 6-30-68, of Arizona Game and Fish Dept. Wildlife Views 16 (1): 8.

——— and D. Berlinski
1959. Waterfowl hunt data collection. Ariz. Game and Fish Dept. Completion Rep. proj. W-70-R-7, work plan 2, job no. 1: 1–6.

Gabrielson, I. N.
1941 (?) What is behind the waterfowl regulations? In report submitted by Mr. Clark to the Senate of the 77th Congress, 1st Session. 34 pp.
Contains map of flyways; map of breeding range of Mallard includes Arizona.

Gallizioli, S.
1951. Graham Mountains special Turkey-Squirrel hunt. Ariz. Game and Fish Comm. Spec. Rep. proj. 53-R-2, work plan 2, job no. 7: 1–4.

———
1951. Annual midsummer Quail survey. Ariz. Game and Fish Comm. Completion Rep. proj. 53-R-2, work plan 3, job no. 7: 1–12.

———
1952. 1951 sectional Quail hunts. Ariz. Game and Fish Comm. Spec. Rep. proj. 53-R-2, work plan 2: 1–6.

———
1952. 1951 Dove hunt. Ariz. Game and Fish Comm. Completion Rep. proj. 53-R-2, work plan 3, job no. 8: 1–3.

———
1952. Pre-season and post-season survey of the Oracle Junction area Quail range. Ariz. Game and Fish Comm. Completion Rep. proj. 53-R-2, work plan 3, job no. 9: 1–8.

———
1953. Annual Quail survey and investigations. Ariz. Game and Fish Dept. Completion Rep. proj. W-53-R-3, work plan 3, job no. 7: 1–19.

———
1953. Dove investigations. Ariz. Game and Fish Dept. Completion Rep. proj. W-53-R-3, work plan 3, job no. 8: 1–25, 1a–9a.

———
1953. Investigation of the Mt. Graham area. (Zone 2, Unit 12). Ariz. Game and Fish Dept. Completion Rep. proj. W-53-R-3, work plan 5, job no. 1: 1–16.

———
1954. Quail hunt information. Ariz. Game and Fish Dept. Completion Rep. proj. W-53-R-4, work plan 2, job no. 8: 1–6.

———
1954. Dove hunt information. Ariz. Game and Fish Dept. Completion Rep. proj. W-53-R-4, work plan 2, job no. 9: 1–13.

———
1954. Investigation of Galiuro Mountains. Ariz. Game and Fish Dept. Completion Rep. proj. W-53-R-4, work plan 5, job no. 1: 1–8.
Mearns' Quail recorded.

———
1955. Quail hunt information. Ariz. Game and Fish Dept. Completion Rep. proj. W-53-R-5, work plan 2, job no. 8: 1–7.

———
1955. Dove hunt information. Ariz. Game and Fish Dept. Completion Rep. proj. W-53-R-5, work plan 2, job no. 9: 1–12.

———

1955. Quail survey and investigations, Gambel's, Scaled and Mearns'. Ariz. Game and Fish Dept. Completion Rep. proj. W-53-R-5, work plan 3, job no. 8: 1–15.

———

1955. Dove investigations, White-winged, Mourning, Inca and Rock. Ariz. Game and Fish Dept. Completion Rep. proj. W-53-R-5, work plan 3, job no. 9: 1–5.

———

1955. Hunting season information on the Mourning and White-winged Doves of Arizona. Proc. 35th Ann. Conf. West. Assoc. State Game and Fish Comms.: 226–235.
Includes data on Whitewing migration.

———

1956. Quail hunt information. Ariz. Game and Fish Dept. Completion Rep. proj. W-53-R-6, work plan 2, job no. 8: 1–11.

———

1956. Dove hunt information. Ariz. Game and Fish Dept. Completion Rep. proj. W-53-R-6, work plan 2, job no. 9: 1–14.

———

1956. That controversial 1955 Quail season. Wildlife News 3 (2): 8–9.
The low Quail population of 1955 was caused by a scarcity of green feed following an unusually dry winter.

———

1959. Arizona game research. Wildlife Views 6 (4): 8–12.
A summary.

———

1960. About Quail and Quail hunting. Contr. Fed. Aid Proj. W-78-R, Ariz. Game and Fish Dept. 7 pp., 8 figs.
Stresses the value of research.

———

1961. The current status and management of the Mourning Dove in the western management unit. Trans. 26th N. A. Wildlife and Nat. Resources Conf. March 6, 7, and 8, 1961: 395–405.
Contains data on abundance and hunting pressure in Arizona.

———

1962. Water and Gambel Quail. Wildlife Views 9 (3): 16–19.

———

1963. Research section. [In] Annual Report, July 1, 1961 to June 30, 1962, of Ariz. Game and Fish Dept. Wildlife Views 10 (1): 14–16.
Contains a description of census methods.

———

1964. Results of a brief investigation of the Masked Bobwhite in Sonora, Mexico. Ariz. Game and Fish Dept. Spec. Rep. 15 pp.
Includes a summary of efforts to restock this bird in Arizona.

———

1964. Research. [In] Annual Report, July 1, 1962 to June 30, 1963, of Ariz. Game and Fish Dept. Wildlife Views 11 (1): 14–16.
Turkeys can be trapped in midsummer.

Gallizioli, S.
1965. Quail research in Arizona. Contr. Fed. Aid Proj. W-78-R, Ariz. Game and Fish Dept. 12 pp., 11 figs.
Brief summary of 15 years of research.

———
1965. Research. [In] Annual Report, July 1, 1963 to June 30, 1964, of Ariz. Game and Fish Dept. Wildlife Views 12 (1): 14–18.
Discusses the value of the Quail call count.

———
1966. Research. [In] Annual Report, July 1, 1964 to June 30, 1965, of Ariz. Game and Fish Dept. Wildlife Views 13 (1): 11–16.
Includes data on Quail call counts.

———
1967. Research. [In] Annual Report, July 1, 1965 to June 30, 1966, of Ariz. Game and Fish Dept. Wildlife Views 14 (1): 48–55.
A summary.

———
1967. Sex and age differential vulnerability to trapping and shooting in Gambel's Quail. [In] Proc. 47th Ann. Conf. W. Assoc. State Game and Fish Comms.: 262–271.
The unbalanced sex ratio in shot and trapped samples is due in part to biased data.

———
1967. Mearns' Quail. Arizona's finest upland game bird. Wildlife Views 14 (6): 4–7.
The population is directly related to the density of the grass cover.

——— and G. Day
1954. Quail survey and investigations. Ariz. Game and Fish Dept. Completion Rep. proj. W-53-R-4, work plan 3, job no. 8: 1–18.

———, S. Levy, and J. Levy
1967. Can the Masked Bobwhite be saved from extinction? Aud. Field Notes 21: 571–575.
They are hopeful that it can be reestablished in Arizona.

——— and R. Smith
1960. Gambel Quail and Cottontail Rabbit population trend techniques. Ariz. Game and Fish Dept. Completion Rep. proj. W-78-R-4, work plan 1, job nos. 2 and 3: 1–10.

——— and E. L. Webb
1958. The influence of hunting upon Quail populations. Ariz. Game and Fish Dept. Completion Rep. proj. W-78-R-3, work plan 4, job no. 3: 1–16.

——— and E. L. Webb
1960. The influence of hunting upon Quail populations. Ariz. Game and Fish Dept. Completion Rep. proj. W-78-R-4, work plan 4, job no. 3: 1–15.

——— and P. M. Webb
1954. Dove investigations. Ariz. Game and Fish Dept. Completion Rep. proj. W-53-R-4, work plan 3, job no. 9: 1–6.

Glover, F. A. and J. D. Smith
1963. Waterfowl status report, 1963. U.S. Fish and Wildlife Serv. Spec. Sci. Rept. Wildl. 75: 1–178.
The statistical report includes Arizona.

Gorsuch, D. M.
1932. Report of Gambel Quail investigation. Ariz. Wild Life 4 (3): 3–7.
Important data on behavior.

———
1939. Game Commission gets Quail survey report. Ariz. Wildlife and Sportsman 1(9): 4–6.

Greenway, J. C., Jr.
1958. Extinct and vanishing birds of the world. Amer. Comm. for Int. Wildl. Protection, N.Y. Spec. Publ. 13. 518 pp.
Pp. 203–204, account of Masked Bobwhite Quail in Arizona.

Griner, L., et al.
194–? Investigational report and proposed Quail management plan for Cochise County, Arizona. (An activity of Ariz. Fed. Aid Proj. 9-R) Ariz. Game and Fish Comm. 19 pp.

Hall, J. M.
1950. Turkey sex, age ratio and population trend survey. Ariz. Game and Fish Comm. Completion Rep. proj. 25-R-3, job no. 1: 1–8.

———
1952. Turkey hunt information. Ariz. Game and Fish Comm. Completion Rep. proj. 53-R-1, work plan 2, job no. 7: 1–3.

———
1952. Turkey hunt information. Ariz. Game and Fish Comm. Completion Rep. proj. 53-R-2, work plan 2, job no. 7: 1–5.

Hewitt, O. H., ed.
1967. The Wild Turkey and its management. The Wildlife Soc. Wash. D.C. 589 pp.
A comprehensive account. Chapter 16 on Merriam's Turkey is by D. MacDonald and R. A. Jantzen.

Hornaday, W. T.
1913. Our vanishing wildlife. New York Zool. Soc. New York: 1–411.
Colinus ridgwayi is considered to be "totally exterminated" in Arizona.

Jackson, A. W. (W. G. Swank, writer)
1955. Investigation of the status of other species of wildlife as related to Beaver. Ariz. Game and Fish Dept. Completion Rep. proj. W-69-R-2, work plan 7, job no. 1: 1–3.
Discusses its influence on Ducks and Turkey.

Jackson, E. B. and P. A. Tilt

1965. Safflower can be harvested before the Doves eat it all. Prog. Agric. Ariz. 17 (5): 6–7.

By harvesting early, the depredations of the White-winged Doves can be decreased.

Jantzen, R. A.

1954. Turkey hunt information. Ariz. Game and Fish Dept. Completion Rep. proj. W-53-R-4, work plan 2, job no. 7: 1–8.

———

1955. Report on Turkey questionnaire cards concerning the 1954 season. Ariz. Game and Fish Dept. Spec. Rep. proj. W-53-R-5, work plan 2, job no. 7: 1–4.

———

1955. Turkey hunt information. Ariz. Game and Fish Dept. Completion Rep. proj. W-53-R-5, work plan 2, job no. 7: 1–15.

———

1955. Turkey survey. Ariz. Game and Fish Dept. Completion Rep. proj. W-53-R-5, work plan 3, job no. 7: 1–11.

———

1956. Report on Turkey questionnaire cards concerning the 1955 season. Ariz. Game and Fish Dept. Spec. Rep. proj. W-53-R-6, work plan 2, job no. 7: 1–4.

———

1956. Turkey hunt information. Ariz. Game and Fish Dept. Completion Rep. proj. W-53-R-6, work plan 2, job no. 7: 1–9.

———

1956. Turkey survey. Ariz. Game and Fish Dept. Completion Rep. proj. W-53-R-6, work plan 3, job no. 7: 1–12.

———

1957. Turkey hunt information. Ariz. Game and Fish Dept. Completion Rep. proj. W-53-R-7, work plan 2, job no. 7: 1–11.

———

1957. Turkey survey. Ariz. Game and Fish Dept. Completion Rep. proj. W-53-R-7, work plan 3, job no. 7: 1–16.

———

1957. Turkey survey. Ariz. Game and Fish Dept. Completion Rep. proj. W-53-R-8, work plan 3, job no. 7: 1–10.

———

1958. Turkey hunt information. Ariz. Game and Fish Dept. Completion Rep. proj. W-53-R-8, work plan 2, job no. 7: 1–11.

———

1961. Turkey management information. Ariz. Game and Fish Dept. Completion Rep. proj. W-53-R-11, work plan 2, job no. 5: 1–5.

———
1964. Big game management. [In] Annual Report, July 1, 1962 to June 30, 1963, of Ariz. Game and Fish Dept. Wildlife Views 11(1): 5–8.
Turkeys are banded.

———
1965. Game management. [In] Annual Report, July 1, 1963 to June 30, 1964, of Ariz. Game and Fish Dept. Wildlife Views 12 (1): 4–5.
Mentions that 3000 doves were killed by a pesticide in Arizona.

Jantzen, R., J. Russo, and T. Knipe
1957. Game habitat improvement recommendations. Ariz. Game and Fish Dept. Completion Rep. proj. W-53-R-7, work plan 7, job no. 1: 1–16.

Judd, S. D.
1905. The Bobwhite and other Quails of the United States in their economic relations. U.S. Dept. Agric. Bur. Biol. Surv. Bull. 21: 1–66.
All Arizona species are discussed.

———
1905. The Grouse and Wild Turkeys of the United States, and their economic value. U.S. Dept. Agric. Biol. Surv. Bull. 24: 1–55.
Brief notes on the Turkey in Arizona.

Kalmbach, E. R.
1939. The Crow in its relation to agriculture. U.S. Dept. Agric. Farmers' Bull. 1102: 1–22.
It is believed to have little economic significance in Arizona.

Kiel, W. H., Jr.
1959. Mourning Dove management units, a progress report. U.S. Fish and Wildlife Serv. Spec. Sci. Rep. Wildl. 42: 1–24.
Contains banding data from Arizona.

———
1960. Mourning Dove newsletter—1960. U.S. Dept. Int. Bur. Sport Fisheries and Wildl.: 1–45.
Includes report on status and studies in Arizona.

Kimball, T. L.
1941. Chukars establish in Arizona. Ariz. Wildlife and Sportsman 4 (1): 8.
Near Maricopa.

———
1946. One man's view of Quail management. Ariz. Wildlife and Sportsman 7 (11): 7–8, 16–17.

———
1948. Gambel Quail management in Arizona. Proc. 28th Ann. Conf. West. Assoc. State Game and Fish Comms. June 2, 3, 4, 1948: 122–124.
Young Quail have a 49 percent mortality over a four month period.

———
1949. Southwestern game management problems. Proc. 29th Ann. Conf. West. Assoc. State Game and Fish Comms. June 14, 15, 16, 1949: 57–60.

Knipe, T.
1951. Classification, population trend and game distribution surveys by management units. Ariz. Game and Fish Comm. Completion Rep. proj. 53-R, work plan 3, job nos. 4 and 5: 1–20.

———
1967. Region VI. [In] Annual Report, July 1, 1965 to June 30, 1966, of Ariz. Game and Fish Dept. Wildlife Views 14 (1): 45–46.
Hunters bagged 910 Mearns' Quail.

Kufeld, R. C.
1962. Turkey management information. Ariz. Game and Fish Dept. Completion Rep. proj. W-53-R-12, work plan 2, job no. 5: 1–4.

———
1962. Quail survey and hunt information. Ariz. Game and Fish Dept. Completion Rep. proj. W-53-R-12, work plan 3, job no. 1: 1–5.

———
1962. The Gila River Dove habitat. Wildlife Views 9 (4): 4–5.

———
1963. Small game. [In] Annual Report, July 1, 1961 to June 30, 1962, of Ariz. Game and Fish Dept. Wildlife Views 10 (1): 9–10.
Data on banding of Doves.

———
1963. Quail survey and hunt information. Ariz. Game and Fish Dept. Completion Rep. proj. W-53-R-13, work plan 3, job no. 1: 1–5.

———
1964. Quail survey and hunt information. Ariz. Game and Fish Dept. Completion Rep. proj. W-53-R-14, work plan 3, job no. 1: 1–5.

———
1964. Dove management information. Ariz. Game and Fish Dept. Completion Rep. proj. W-53-R-14, work plan 3, job no. 2: 1–9.

———
1964. Small game. [In] Annual Report, July 1, 1962 to June 30, 1963, of Ariz. Game and Fish Dept. Wildlife Views 11 (1): 8–10.
Data on Doves and Quail.

———
1965. Small game. [In] Annual Report, July 1, 1963 to June 30, 1964, of Ariz. Game and Fish Dept. Wildlife Views 12 (1): 10–11.
Includes data on Quail populations and banding of Doves.

———
1966. Small game. [In] Annual Report, July 1, 1964 to June 30, 1965, of Ariz. Game and Fish Dept. Wildlife Views 13 (1): 8–9.
Data on Doves, Quail, and Blue Grouse.

——— and L. Fitzhugh
1962. Dove survey and hunt information. Ariz. Game and Fish Dept. Completion Rep. proj. W-53-R-12, work plan 3, job no. 2: 1–7.

[Lawson, L. L.]
1949. Whitewing Dove study. Ariz. Game and Fish Comm. Completion Rep. proj. 40-R, job no. 5: 1–7.
Data on nesting.

———
1949. Annual midsummer Quail survey. Ariz. Game and Fish Comm. Completion Rep. proj. 40-R, job no. 6: 1–5.

Lawson, L. L.
1950. Quail research and investigation in southern Arizona. Ariz. Game and Fish Comm. Completion Rep. proj. 40-R-2, work plan 1, job no. 1: 1–4.

———
1950. White-winged and Mourning Dove investigations. Ariz. Game and Fish Comm. Completion Rep. F. A. proj. 40-R-2, job no. 1: 1–11.

———
1951. Special investigations on the status of Pheasant and Chukar Partridge in Arizona. Ariz. Game and Fish Comm. Completion Rep. proj. 40-R-2, work plan 2, job no. 3: 1–5.
Pheasant successful only in Springerville area.

Levy, S. H., J. J. Levy, and R. A. Bishop
1966. Use of tape recorded female Quail calls during the breeding season. Jour. Wildl. Mgmt. 30: 426–428.
In southeastern Arizona, male Harlequin, Gambel's and Scaled Quail responded to the recorded calls.

Ligon, J. S.
1946. History and management of Merriam's Wild Turkey. Univ. New Mexico Publ. Biol. 1: 1–84.
Includes maps of past and present range in Arizona.

Linduska, J. P.
1967. Game never had it so good. Wildlife Views 14 (5): 20–21.
Includes photograph of Canada Geese flock introduced at White Mountain Lake, Arizona.

Manes, R. R.
1968. The fatal future. Death dirge for Dove. Wildlife Views 15 (3): 22–25.
The Corps of Engineers is destroying the Dove habitat along thc Gila River.

———
1968. The fatal future. Threatened species doomed. Wildlife Views 15 (4): 4–9.
Plans of the Corps of Engineers to destroy the habitat along the San Pedro River.

———
1968. The fatal future. Part III. The Lower Colorado River. Wildlife Views 15 (5): 22–25.
Losses of wildlife predicted if the Corps of Engineers remove the plants along the river.

Martin, S. C.
1966. Will you see any game today? Prog. Agric. Ariz. 18 (4): 30–31.
Brief account of surveys on the Santa Rita Experimental Range, Arizona.

Martin, W. M.
1967. Region III. [In] Annual Report, July 1, 1965 to June 30, 1966, of Ariz. Game and Fish Dept. Wildlife Views 14 (1): 39–41.
Quail were abundant near permanent water.

McAtee, W. L.
1945. The Ring-necked Pheasant and its management in North America. Amer. Wildl. Inst. 320 pp.
The map of its range in Arizona in 1941 included the Gila and Salt rivers. It did not become established.

McCulloch, C. Y.
1960. The influence on carrying capacity of experimental water conservation measures. Ariz. Game and Fish Dept. Completion Rep. proj. W-78-R-4, work plan 5, job no. 7: 1–8.

McMurray, F. B.
1946. Where were the waterfowl on Lake Laguna this year? Ariz. Wildlife and Sportsman 7 (12): 7, 14.
Discusses reasons for the decrease.

Moody, R. J.
1936. Effect of insecticides on game birds. Ariz. Wildlife and Sportsman 7 (1): 11.
He believes they cause little injury to birds.

Murie, A.
1946. The Merriam Turkey on the San Carlos Indian Reservation. Jour. Wildl. Mgmt. 10: 329–333.
Data on food habits.

Neff, J. A.
1940. Range, population, and game status of the Western White-winged Dove in Arizona. Jour. Wildl. Mgmt. 4: 117–127.

———
1952. Inventory of Band-tailed Pigeon populations in Arizona, Colorado, and New Mexico. U.S. Dept. Int. Fish and Wildlife Serv. Wildl. Research Lab. Denver: 1–26.

Nichol, A. A.
1939. Pittman-Robertson progress report. Ariz. Wildlife and Sportsman 1 (12): 5–6, 11–12.
Includes Quail, Turkey, and Pheasant reports.

———
1940. Monthly report P-R projects. Ariz. Wildlife and Sportsman 2 (1): 7–9.
Includes Quail survey in Yuma County.

Nichuis, C. C.
1950. Duck detectives. Ariz. Wildlife and Sportsman 11 (12): 5, 14.
Waterfowl banding in Arizona.

Packard, L.
1967. Region II. [In] Annual Report, July 1, 1965 to June 30, 1966, of Ariz. Game and Fish Dept. Wildlife Views 14 (1): 38–39.
The waterfowl hunt was good in the Chino Valley and Anderson Mesa areas.

———
1968. Region II—Flagstaff. [In] Annual Report, July 1, 1966 to June 30, 1967, of Ariz. Game and Fish Dept. Wildlife Views 15 (1): 32.
The Chukar Partridge is extending its range.

Powell, L. E.
1954. Investigation of Bagdad-Kingman area. Ariz. Game and Fish Dept. Completion Rep. proj. W-53-R-4, work plan 5, job no. 11: 1–17.

Reed, J. J.
1958. Investigation of the Mazatzal Mountains. Ariz. Game and Fish Dept. proj. W-53-R-8, work plan 5, job no. 22: 1–4.
Includes brief mention of game birds.

———
1963. A report on those Rio Grande Turkeys. Wildlife Views 10 (4): 16–17.
Released near Ruby, Arizona.

Reeves, H. M., A. D. Geis, and F. C. Kniffen
1968. Mourning Dove capture and banding. U.S. Dept. Int. Bur. Sport Fisheries and Wildlife. Spec. Sci. Rep. Wildl. 117: 1–63.
Includes trapping at Tucson, Arizona, p. 36.

Reeves, R. H.
1950. Turkey winter range investigation. Ariz. Game and Fish Comm. Completion Rep. proj. 49-R-1, job no. 1: 1–15.

———
1950. Turkey summer range investigation. Ariz. Game and Fish Comm. Completion Rep. proj. 49-R, job no. 2: 1–26.

———
1950. Turkey population trend and age and sex ratio survey development. Ariz. Game and Fish Comm. Completion Rep. proj. 49-R, job no. 3: 1–5.

———
1951. Turkey winter range investigation. Ariz. Game and Fish Comm. Completion Rep. proj. 49-R-2, job no. 1: 1–22.

———
1951. Turkey summer range investigation. Ariz. Game and Fish Comm. Completion Rep. proj. 49-R-2, job no. 2: 1–17.

———
1951. Turkey population trend and age and sex ratio survey development. Ariz. Game and Fish Comm. Completion Rep. proj. 49-R-2, job no. 3: 1–5.

———
1952. Wild Turkey management in Arizona. Proc. 32nd Ann. Conf. West. Assoc. State Game and Fish Comms. June 15, 16, 17, 1952: 106–109.
Mortality, food habits, predators, and populations.

Reeves, R. H.
1953. Turkey hunt information. Ariz. Game and Fish Dept. Completion Rep. proj. W-53-R-3, work plan 2, job no. 7: 1–7.

———
1954. Turkey trapping and banding. Ariz. Game and Fish Dept. Spec. Rep. proj. W-53-R-4, work plan 3, job no. 7: 1–9.

——— and W. G. Swank
1955. Food habits of Merriam's Turkey. Ariz. Game and Fish Dept. Completion Rep. proj. W-49-R-3, work plan 1, job no. 3: 1–15.

Russo, J. P.
1954. Mittry Lake investigation. Ariz. Game and Fish Dept. Completion Rep. proj. W-53-R-4, work plan 5, job no. 12: 1–9.
Contains a list of migratory waterfowl.

———
1956. Investigation of Ajo vicinity. Ariz. Game and Fish Dept. Completion Rep. proj. W-53-R-6, work plan 5, job no. 21: 1–4.

———
1956. The desert bighorn sheep in Arizona. Ariz. Game and Fish Dept. 153 pp.
Contains a discussion of Golden Eagle predation.

———
1958. Kaibab Turkey transplant. Proc. 38th Ann. Conf. West. Assoc. State Game and Fish Comms. Pp. 175–178.
It was successful.

———
1965. Big game management. [In] Annual Report, July 1, 1963 to June 30, 1964, of Ariz. Game and Fish Dept. Wildlife Views 12 (1): 5, 7, 9.
Includes a summary of the Turkey harvest.

———
1966. Big game management. [In] Annual Report, July 1, 1964 to June 30, 1965, of Ariz. Game and Fish Dept. Wildlife Views 13 (1): 5–7.
Includes a summary of the Turkey harvest.

———
1967. Big game management. [In] Annual Report, July 1, 1965 to June 30, 1966, of Ariz. Game and Fish Dept. Wildlife Views 14 (1): 5–6.
Includes summary of Turkey hunts.

———
1968. Big game. [In] Annual Report, July 1, 1966 to June 30, 1967, of Ariz. Game and Fish Dept. Wildlife Views 15 (1): 6.
No permit restrictions on the fall Turkey season hunt.

———
1969. Big game. [In] Annual Report, July 1, 1967 to June 30, 1968, of Ariz. Game and Fish Dept. Wildlife Views 16 (1): 5–6.

Schimmel, B. [= R.]
1967. Return of the Bandtails. Wildlife Views 14 (5): 22–26.
Brief notes on banding, with map of range in Arizona.

———

1968. The fatal future. Part IV. The Santa Cruz River. Wildlife Views 15 (6): 24–27.

Discusses the disaster to wildlife if the streamside vegetation is removed.

[Schimmel, R., ed.]

1970. 40 years of progress. Wildlife Views 17 (1): Game management, 16–26, 38–42; research, 35–37, 60–61.

Schorger, A. W.

1966. The wild turkey. Its history and domestication. Univ. Okla. Press, Norman: 1–625.

Good Arizona history.

Shaw, H.

1961. The influence of salt cedar on White-winged Doves in the Gila Valley. Ariz. Game and Fish Dept. Spec. Rep.: 1–9.

——— and J. Jett

1959. Mourning Dove and White-winged Dove nesting in the Gila River bottom between Gillespie Dam and the junction of the Salt and Gila Rivers, Maricopa County, Arizona. Ariz. Game and Fish Dept. Spec. Rep. proj. W-53-R-10, work plan 3, job no. 2: 1–6, 4 maps.

Sizer, J. W.

1963. The small game "numbers game." Wildlife Views 10 (5): 4–7.

A discussion of natural controls of animal populations.

[Sizer, B. (=J. W.)]

1965. Editorial: where are these conservationists? Wildlife Views 12 (2): 3.

Opposes the destruction of the Salt and Gila River habitat of the White-winged Dove by the Army Engineers.

Sizer, B. (=J. W.)

1965. Gila River wildlife—Habitat or epitaph? Wildlife Views 12(2): 4–9.

A discussion of the proposed channel in the Phoenix region and its effect on the White-winged Dove population.

[Sizer, J. W.]

1967. What's going on here? Wildlife Views 14 (2): 24.

Photo of White-winged Pheasants being released in the Safford Valley, Arizona.

Smith, D. F.

1966. Region IV. [In] Annual Report, July 1, 1964 to June 30, 1965, of Ariz. Game and Fish Dept. Wildlife Views 13 (1): 31–33.

Comments on the destruction of habitat by channelization of the Colorado River.

Smith, R. H.

[1960?] Turkey population trend techniques. Ariz. Game and Fish Dept. Completion Rep. proj. W-78-R-4, work plan 1, job no. 5: 1–7.

———

1961. Big game hunt questionnaire survey. Ariz. Game and Fish Dept. Spec. Rep. proj. W-53-R-11, work plan 2, job nos. 1–5: 1–10.

Includes Turkey.

Smith, R. H.

1961. Results of pilot work on a small game hunter take survey. Ariz. Game and Fish Dept. Suppl. Rep. proj. W-53-R-11, work plan 3, job nos. 1, 2, 3, 4: 1–7.

———

1961. Turkey population trend techniques. Ariz. Game and Fish Dept. Completion Rep. proj. W-78-R-5, work plan 1, job no. 5: 1–9.

———

1962. Small game hunter take survey—1961. Ariz. Game and Fish Dept. Suppl. Rep. proj. W-53-R-12, work plan 3, job nos. 1, 2, 3, 4: 1–12.

———

1964. Predicting Gambel Quail hunter success from call counts. Proc. 3rd Ann. Meeting Wildl. Soc. New Mex.-Ariz. Sec. Feb. 7, 8, 1964: 43–49.

Includes data on rainfall and reproduction in Arizona.

———

1966. Hunt information. [In] Annual Report, July 1, 1964 to June 30, 1965, of Arizona Game and Fish Dept. Wildlife Views 13 (1): 9–10.

The summary includes harvest of Doves, Quail, and Turkey, with hunter success.

———

1968. Game research. [In] Annual Report, July 1, 1966 to June 30, 1967, of Ariz. Game and Fish Dept. Wildlife Views 15 (1): 26–29.

Includes Turkey studies.

———

1969. Game research. [In] Annual Report, July 1, 1967 to June 30, 1968, of Ariz. Game and Fish Dept. Wildlife Views 16 (1): 20–25.

Studies of Mearns' Quail and White-winged Dove continued.

——— and S. Gallizioli

1958. Gambel Quail population trend techniques. Ariz. Game and Fish Dept. Completion Rep. proj. W-78-R-3, work plan 1, job no. 3: 1–15.

——— and S. Gallizioli

1963. Gambel Quail population trend techniques. Ariz. Game and Fish Dept. Completion Rep. proj. W-78-R-7, work plan 1, job no. 3: 1–22.

——— and S. Gallizioli

1965. Predicting hunter success by means of a spring call count of Gambel Quail. Jour. Wildl. Mgmt. 29: 806–813.

"The average absolute deviation of predicted success from actual success was only 12 percent."

Sowls, L. K.

1956. Wildlife conservation through cooperation. Univ. Ariz. Press. 32 pp.

Summaries of Dove and Quail research in Arizona are included.

———

1960. Arizona Cooperative Wildlife Research Unit. Wildlife Views 7 (1): 40–41.

History and summary of work.

———

1960. Results of a banding study of Gambel's Quail in southern Arizona. Jour. Wildl. Mgmt. 24: 185–190.

Near Oracle.

———

1961. Arizona Cooperative Wildlife Research Unit. [In] Annual Report, July 1, 1959 to June 30, 1960, of Ariz. Game and Fish Dept. Wildlife Views 8 (1): 21–23.

Brief report on Gambel's Quail studies.

———

1962. Arizona Cooperative Wildlife Research Unit. [In] Annual Report, July 1, 1960 to June 30, 1961, of Ariz. Game and Fish Dept. Wildlife Views 9 (1): 15–18.

Includes Gambel's Quail studies.

———

1963. Arizona Cooperative Wildlife Research Unit. [In] Annual Report, July 1, 1961 to June 30, 1962, of Ariz. Game and Fish Dept. Wildlife Views 10 (1): 17–19.

Data on banding of Doves.

———

1964. Arizona Cooperative Wildlife Research Unit. [In] Annual Report, July 1, 1962 to June 30, 1963, of Ariz. Game and Fish Dept. Wildlife Views 11 (1): 20–23.

Brief report on studies of Mearns' Quail, Mourning Doves, and Masked Bobwhite.

———

1965. Arizona Cooperative Wildlife Research Unit. [In] Annual Report, July 1, 1963 to June 30, 1964, of Ariz. Game and Fish Dept. Wildlife Views 12 (1): 19–22.

Includes a report on research on Mourning Doves and Quail.

———

1966. Arizona Cooperative Wildlife Research Unit. [In] Annual Report, July 1, 1964 to June 30, 1965, of Ariz. Game and Fish Dept. Wildlife Views 13 (1): 17–20.

Includes a report on discovery of the disease trichomoniasis in Doves.

———

1967. Cooperative Wildlife Research Unit. [In] Annual Report, July 1, 1965 to June 30, 1966, of Ariz. Game and Fish Dept. Wildlife Views 14 (1): 12–14.

Includes summary of research on game birds.

———

1968. Cooperative Wildlife Research Unit. [In] Annual Report, July 1, 1966 to June 30, 1967, of Ariz. Game and Fish Dept. Wildlife Views 15 (1): 11–13.

Brief summary of research.

Sowls, L. K.
1969. Cooperative Wildlife Research Unit. [In] Annual Report, July 1, 1967 to June 30, 1968, of Ariz. Game and Fish Dept. Wildlife Views 16 (1): 10–11.
Studies of Doves, Pigeons, and Blue Grouse continued.

——— and L. A. Greenwalt
1956. Large traps for catching Quail. Jour. Wildl. Mgmt. 20: 215–216.
Experiences with traps near Oracle, Arizona.

Stair, J. L.
1956. Dove investigations. Ariz. Game and Fish Dept. Completion Rep. proj. W-53-R-6, work plan 3, job no. 9: 1–11.

———
1957. Quail survey and investigations. Ariz. Game and Fish Dept. Completion Rep. proj. W-53-R-6, work plan 3, job no. 8: 1–10.

———
1957. Quail survey and investigations. Ariz. Game and Fish Dept. Completion Rep. proj. W-53-R-7, work plan 3, job no. 8: 1–9.

———
1957. Dove investigations. Ariz. Game and Fish Dept. Completion Rep. proj. W-53-R-7, work plan 3, job no. 9: 1–32.

———
1958. What is your story, Mrs. Dove? Wildlife Views 5 (3): 12–15, 28–29.
Banding data on Mourning and White-winged Doves in Arizona.

———
1958. Quail hunt and survey information. Ariz. Game and Fish Dept. Completion Rep. proj. W-53-R-8, work plans 2 and 3, job no. 8: 1–13.

———
1958. Dove hunt information and Dove investigations. Ariz. Game and Fish Dept. Completion Rep. proj. W-53-R-8, work plans 2 and 3, job no. 9: 1–36.

———
1958. Determination of small game and non-game kill. Ariz. Game and Fish Dept. Completion Rep. proj. FW-11-R-1, job no. 2: 1–27.

———
1958. Dove management and what it means to Arizona. Proc. 38th Ann. Conf. West. Assoc. State Game and Fish Comms.: 194–196.
"Doves are abundant in Arizona, not particularly due to management. . . ."

———
1959. Quail hunt and survey information. Ariz. Game and Fish Dept. Completion Rep. proj. W-53-R-9, work plans 2 and 3, job no. 8: 1–10.

———
1959. Dove hunt information and Dove investigations. Ariz. Game and Fish Dept. Completion Rep. proj. W-53-R-9, work plans 2 and 3, job no. 9: 1–25.

———

1959. White-winged Dove die-off at Maricopa. Ariz. Game and Fish Dept. Spec. Rep. proj. W-53-R-9, work plan 3, job no. 9: 1–8.

———

1959. Investigation of the Gila Bend area. Ariz. Game and Fish Dept. Completion Rep. proj. W-53-R-9, work plan 5, job no. 26: 1–8.

Includes Doves.

———

1960. Quail hunt and survey information. Ariz. Game and Fish Dept. Completion Rep. proj. W-53-R-10, work plan 3, job no. 1: 1–15.

———

1960. Dove management information. Ariz. Game and Fish Dept. Completion Rep. proj. W-53-R-10, work plan 3, job no. 2: 1–12.

———

1960. Wildlife and the farmers. Wildlife Views 7 (4): 13–17.

A discussion of the effect of farming on Quail and Doves.

———

1961. Small game. [In] Annual Report, July 1, 1959 to June 30, 1960, of Ariz. Game and Fish Dept. Wildlife Views 8 (1): 3–4.

Includes data on Doves and Quail.

———

1961. Quail management information. Ariz. Game and Fish Dept. Completion Rep. proj. W-53-R-11, work plan 3, job no. 1: 1–6.

———

1961. Dove management information. Ariz. Game and Fish Dept. Completion Rep. proj. W-53-R-11, work plan 3, job no. 2: 1–17.

———

1961. Arizona's first Mearns' Quail hunt. Proc. 41st Ann. Conf. West. Assoc. State Game and Fish Comms.: 172–175.

Stanley, R.

1952. Quacker-Honker comeback. Ariz. Wildlife and Sportsman 23 (12): 30–39.

A report on the Arizona populations and distribution.

Swank, W. G.

1954. Working for a bigger Turkey crop. Ariz. Wildlife and Sportsman 25 (2): 26–31.

A report on trapping and habits of Turkeys, with photographs.

——— and S. Gallizioli

1954. The influence of hunting and of rainfall upon Gambel's Quail populations. Trans. 19th N. A. Wildl. Conf. March 8, 9, and 10, 1954: 283–297.

Winter rainfall limits abundance. Good rains produce spring feed and more Quail.

——— and S. Gallizioli

1958. The effects of hunting on Gambel Quail populations. Trans. 23rd N. A. Wildl. Conf. March 3, 4, 5, 1958: 305–319.

Swank, W. G., et al.
1954. Turkey survey. Ariz. Game and Fish Dept. Completion Rep. proj. W-53-R-4, work plan 3, job no. 7: 1–12.

[Tallon, J., ed.]
1966. "What's going on here?" Wildlife Views 13 (3): 24.
Brief report that White-winged Pheasants have been introduced into the Phoenix area.

———
1966. "What's going on here?" Wildlife Views 13 (4): 24.
Photograph of Canada Geese goslings that are to be released in the White Mountains, Arizona.

Tomlinson, R. E.
1968. Reward banding to determine reporting rate of recovered Mourning Dove bands. Jour. Wildl. Mgmt. 32: 6–11.
Lists one Arizona bird recovered in Mexico.

——— and D. E. Brown
1970. Our Bobwhites come home. Wildlife Views 17 (3): 4–11.
Report on recent introductions in Arizona.

Towell, W. E.
1958. Report of endangered species of wildlife committee. [In] 48th Conv. Int. Assoc. Game, Fish and Conservation Comms.: 35–44.
Mearns' Quail, p. 36: habitat improvement required in Arizona.

Vincent, J., compiler.
1966. [Rare and endangered species] Aves. Survival Service Commission Red Data Book, vol. 2. Int. Union Cons. Nature and Nat. Resources. 275 pp.
Lists *Anas diazi* and *Colinus virginianus ridgwayi* in Arizona.

Vorhies, C. T.
1935. A blot on G.P.A. activities. Ariz. Wild Life 6 (9): 1.
Condemns a campaign offering bounties on Hawks, Crows, and Roadrunners.

———
1942. The White-winged Dove as a game bird. Ariz. Wildlife and Sportsman 4 (4): 6–7.
Objects to hunting during their breeding season.

———
1943. Game birds in the University collection. Ariz. Wildlife and Sportsman 5 (3): 8.

Wagner, R. A.
1952. Waterfowl resources of the Colorado River. Ariz. Game and Fish Comm. Proj. 60-R, job no. 1: 1–5.
Data on population density.

———
1952. Small game resources of the Colorado River. Ariz. Game and Fish Comm. Proj. 60-R, job no. 2: 1–9.
Data on population density of Quail and Doves.

Walker, L. W.
1964. Return of the Masked Bobwhite. Zoonooz 37 (1): 10–15.
A summary of the preliminary attempt to establish this species in the vicinity of Tucson, Arizona.

———
1964. Return of the Masked Bobwhite. Ariz.-Son. Desert Museum, Spec. Bull. 7 pp.
Describes an attempt to reestablish the species in Arizona.

Wallmo, C. O.
1951. Range, distribution and wildlife inventory of species on Fort Huachuca area. Ariz. Game and Fish Comm. Proj. 46-R-1, job no. 3: 1–30.
Includes a report on Turkey and various species of Quail.

———
1951. General wildlife surveys of the Fort Huachuca wildlife area. Ariz. Game and Fish Comm. Proj. 46-R-2, job no. 2: 1–10.
Data on populations and nesting of Band-tailed Pigeons, Turkey, and Quail.

——— and C. McCulloch
1963. Influence on carrying capacity of experimental water conservation measures. Ariz. Game and Fish Dept. Completion Rep. proj. W-78-R-7, work plan 5, job no. 7: 1–12.

Webb, E. L.
1958. Water development evaluation. Proc. 38th Ann. Conf. West. Assoc. State Game and Fish Comms.: 251–255.
Effect on Gambel's Quail populations.

———
1959. The effect of water development on the distribution and abundance of Quail, Deer, and Javelina. Ariz. Game and Fish Dept. Completion Rep. proj. W-78-R-3, work plan 4, job no. 4: 1–10.

———
1960. The effect of water development on the distribution and abundance of Quail and Deer. Ariz. Game and Fish Dept. Completion Rep. proj. W-78-R-4, work plan 4, job no. 4: 1–7.

——— and S. Gallizioli
1963. The effect of water development on the abundance of Quail and Deer. Ariz. Game and Fish Dept. Completion Rep. proj. W-78-R-6, work plan 4, job no. 4: 1–7.

Webb, P. M.
1953. Trapping, tagging, and releasing Quail on the Three Bar area. Ariz. Game and Fish Comm. Completion Rep. proj. W-11-D-13, work plan 4, job no. 2: 1–12.

———
1953. Introduction and study of exotic bird species. Ariz. Game and Fish Dept. Completion Rep. proj. W-58-R-1, work plan 1, job no. 1: 1–4.

———
1954. A study of the introduction, release and survival of Chukar Partridge, Sand Grouse and other game species. Ariz. Game and Fish Dept. Completion Rep. proj. W-58-R-2: 1–5.

Webb, P. M.
1956. A study of the introduction, release and survival of Chukar Partridge, Sand Grouse and other game species. Ariz. Game and Fish Dept. Completion Rep. proj. W-58-R-4, work plan 1, job nos. 2 and 3: 1–5.

———
1957. The introduction of the Grey Francolin in Arizona: a progress report. Proc. 37th Ann. Conf. West. Assoc. State Game and Fish Comms. June 17, 18, 19, 1957: 274–276.
Probably a failure.

———
1959. Distribution, survival and reproduction of introduced species. Ariz. Game and Fish Dept. Completion Rep. proj. W-78-R-3, work plan 6, job no. 2: 1–7.

———
1960. Turkey trapping and transplanting in Arizona. [In] Proc. 40th Ann. Conf. West. Assoc. State Game and Fish Comms. June 20–22, 1960: 182–187.
Chiefly techniques.

———
1962. Trapping and transplanting. [In] Annual Report, July 1, 1960 to June 30, 1961, of Ariz. Game and Fish Dept. Wildlife Views 9 (1): 5.
Data on releases of Merriam Turkey in Arizona.

———
1963. Trapping and transplanting. [In] Annual Report, July 1, 1961 to June 30, 1962, of Ariz. Game and Fish Dept. Wildlife Views 10 (1): 10.
Data on Turkey releases.

———
1964. Trapping and transplanting. [In] Annual Report, July 1, 1962 to June 30, 1963, of Ariz. Game and Fish Dept. Wildlife Views 11 (1): 10.
Data on Turkey and Chukar releases.

———
1967. Small game. [In] Annual Report, July 1, 1965 to June 30, 1966, of Ariz. Game and Fish Dept. Wildlife Views 14 (1): 6–9.
Includes Doves, Quail, Chukar, Blue Grouse, and Pheasant.

———
1968. Small game. [In] Annual Report, July 1, 1966 to June 30, 1967, of Ariz. Game and Fish Dept. Wildlife Views 15 (1): 6–8.
Chiefly hunt summary.

———
1969. Small game. [In] Annual Report, July 1, 1967 to June 30, 1968, of Ariz. Game and Fish Dept. Wildlife Views 16 (1): 6–8.

[Weekes, R. W., ed.]
1927. Pheasant nursery growing. Ariz. Wild Life 1 (1): 6.
Ring-necked Pheasants released on Mt. Lemmon raised broods.

———
1927. What makes Turkeys wild? Ariz. Wild Life 1 (4–5): 9.
Turkeys released in the Chiricahuas became tame.

———

1927. Where are the Ducks? Ariz. Wild Life 1 (1): 12.
Comments on early migration in Arizona in autumn of 1926 and the scarcity later.

White, R. W.
1964. Something new has been added. Blue Grouse. Ariz. Wildlife and Sportsman 35 (9): 19.
First hunt authorized in Arizona. Includes some data on Grouse habits.

———

1967. Region I. [In] Annual Report, July 1, 1965 to June 30, 1966, of Ariz. Game and Fish Dept. Wildlife Views 14 (1): 37–38.
Includes a summary of game bird distribution in the area.

———

1969. Region I—Pinetop. [In] Annual Report, July 1, 1967 to June 30, 1968, of Ariz. Game and Fish Dept. Wildlife Views 16 (1): 26–27.
Includes a report on introduction of Canada Geese.

——— et al.
1965. Turkey management information. Ariz. Game and Fish Dept. Completion Rep. proj. W-53-R-15, work plan 2, job no. 5: 1–11.

Wilbur, S. R.
1967. Live-trapping North American upland game birds. U.S. Dept. Int. Bur. Sport Fisheries and Wildlife. Spec. Sci. Rep.—Wildl. 106: 1–37.
Summaries of Arizona publications on trapping Arizona species.

Wilson, J. R.
1946. Midsummer Quail survey. Ariz. Wildlife and Sportsman 7 (9): 14–15, 23.

Wingfield, [D. H.]
1967. Region VII. [In] Annual Report, July 1, 1965 to June 30, 1966, of Ariz. Game and Fish Dept. Wildlife Views 14 (1): 46–47.
Includes data on Quail distribution in southeastern Arizona.

Wingfield, D.
1968. Region VII—Pima. [In] Annual Report, July 1, 1966 to June 30, 1967, of Ariz. Game and Fish Dept. Wildlife Views 15 (1): 38–39.
Includes a brief report on introduction of the White-winged Pheasant.

———

1969. Region VII—Pima. [In] Annual Report, July 1, 1967 to June 30, 1968, of Ariz. Game and Fish Dept. Wildlife Views 16 (1): 34–35.
White-winged Pheasants are increasing in the Gila Valley.

Wright, J. T.
1952. Investigation and evaluation of wildlife use of water developments. Ariz. Game and Fish Comm. Completion Rep. proj. W-62-R-1, work plan 1, job no. 1: 1–21.

———

1953. Investigation and evaluation of wildlife use of water developments. Ariz. Game and Fish Comm. Completion Rep. proj. W-62-R-2: 1–34.

Wright, J. T.
1959. Desert wildlife. Ariz. Game and Fish Dept. Wildlife Bull. 6: 1–78.
A review of game management work in west-central and southwestern Arizona; data on plants, topography, climate, and fauna; notes on game birds.

——— and D. P. Schadle
1954. The evaluation of water development in wildlife habitat improvement. Ariz. Game and Fish Dept. Completion Rep. proj. W-62-R-3: 1–52.

——— and E. L. Webb
1955. The evaluation of water development in wildlife habitat improvement. Ariz. Game and Fish Dept. Completion Rep. proj. W-62-R-4: 1–51.

8. Theses and Dissertations

Theses and dissertations concerned in some manner with Arizona ornithology are listed here. Most of these were written by students of Arizona universities and are unpublished.

Ambrose, J. E., Jr.

1963. The breeding ecology of *Toxostoma curvirostre* and *T. bendirei* in the vicinity of Tucson, Arizona. M. S. thesis. Univ. Ariz. 40 pp.

T. bendirei is competing unsuccessfully with *T. curvirostre*, and is decreasing.

Arnold, L. W.

1940. An ecological study of the vertebrate animals of the mesquite forest. M.S. thesis. Univ. Ariz. 79 pp.

Records 111 species of birds along the Santa Cruz River south of Tucson.

Balda, R. P.

1967. Ecological relationships of the breeding-birds of the Chiricahua Mountains, Arizona. Ph.D. dissertation. Univ. Ill. 240 pp.

He recorded 103 nesting species in the area.

Bishop, R. A.

1964. The Mearns Quail *(Cyrtonyx montezumae mearnsi)* in southern Arizona. M.S. thesis. Univ. Ariz. 82 pp.

In Canelo Hills and Box Canyon, Santa Rita Mountains.

Braun, E. J.

1969. Metabolism and water balance of the Gila Woodpecker and Gilded Flicker in the Sonoran Desert. Ph.D. dissertation. Univ. Ariz. 92 pp.

"Both species used the saguaro *(Cereus giganteus)* tree-hole to escape the harshest aspects of their desert environment during the summer and winter."

Bulmer, W., Jr.

1966. The breeding biology of the Red-faced Warbler *(Cardellina rubrifrons)*. M.S. thesis. Univ. Ariz. 57 pp.

In the Santa Catalina Mountains, Arizona.

Calder, W. A.

1966. Temperature regulation and respiration in the Roadrunner and the Pigeon. Ph.D. dissertation. Duke Univ. 108 pp.

Arizona Roadrunners were used.

Dalby, S. L.
1969. Prolactin and the orientation of Zugunruhe in the White-crowned Sparrow. M.S. thesis. Univ. Ariz. 23 pp.
Near Tucson, Arizona.

Dickerman, R. W.
1954. An ecological survey of the Three-Bar Game Management Unit located near Roosevelt, Arizona. M.S. thesis. Univ. Ariz. 126 pp.
Lists 51 species of birds.

Elder, J. B.
1953. Utilization of man-made water-holes by wildlife in southern Arizona. M.S. thesis. Univ. Ariz. 114 pp.
Records of 25 species of birds observed drinking.

George, W. G.
1958. The hyoid apparatus of nine-primaried oscinine birds. M.S. thesis. Univ. Ariz. 28 pp.
Arizona specimens used.

———
1961. The taxonomic significance of the tongue musculature of Passerine birds. Ph.D. dissertation. Univ. Ariz. 129 pp.
Arizona specimens used.

Gould, P. J.
1960. Territorial relationships between Cardinals and Pyrrhuloxias. M.S. thesis. Univ. Ariz. 40 pp.
On the San Xavier Indian Reservation, 10 miles south of Tucson. The two species are believed to be congeneric.

Greenwalt, L. A.
1955. The mobility of Gambel's Quail *(Lophortyx gambeli gambeli)* in a desert-grassland-oak-woodland area in southeastern Arizona. M.S. thesis. Univ. Ariz. 39 pp.
Near Oracle, Pinal County.

Gubanich, A. A.
1970. Seasonal variation of neurosecretory material in the neurohypophysis of desert birds. Ph.D. dissertation. Univ. Ariz. 78 pp.

Hannum, C. A.
1941. Nematode parasites of Arizona vertebrates. Ph.D. dissertation. Univ. Wash. 158 pp.
Lists *Ascaridia galli* from *Melopelia leucoptera* and *Subulura gracile* from *Lophortyx gambeli.*

Hensley, M. M.
1951. Ecological relations of the breeding bird population of the desert biome in Arizona. Ph.D. dissertation. Cornell Univ. 243 pp.

Hungerford, C. R.
1960. The factors affecting the breeding of Gambel's Quail (*Lophortyx gambelii gambelii* Gambel) in Arizona. Ph.D. dissertation. Univ. Ariz. 94 pp.
In the Tucson vicinity.

Irby, H. D.
1964. The relationship of calling behavior to Mourning Dove populations and production in southern Arizona. Ph.D. dissertation. Univ. Ariz. 100 pp.

Johnson, R. R.
1960. The biota of Sierra Ancha, Gila County, Arizona. M.S. thesis. Univ. Ariz. 114 pp.
Includes accounts of 127 species of birds.

Knopp, T. B.
1959. Factors affecting the abundance and distribution of Merriam's Turkey *(Meleagris gallopavo merriami)* in southeastern Arizona. M.S. thesis. Univ. Ariz. 58 pp.
In the Graham Mountains.

Krizman, R. D.
1964. The sahuaro tree-hole microenvironment in southern Arizona. I. Winter. M.S. thesis. Univ. Ariz. 35 pp.
Brief notes on hole-inhabiting birds.

Lammers, G. E.
1970. The Late Cenozoic Benson and Curtis Ranch Faunas from the San Pedro Valley, Cochise County, Arizona. Ph.D. dissertation. Univ. Ariz. 193 pp.
Lists 14 species of birds.

LeCount, A. L.
1970. Fall food preferences of Blue Grouse in the White Mountains of Arizona. M.S. thesis. Univ. Ariz. 45 pp.

Ligon, J. D.
1967. The biology of the Elf Owl, *Micrathene whitneyi.* Ph.D. dissertation. Univ. Mich. 158 pp.

Ohmart, R. D.
1969. Physiological and ethological adaptations of the Rufous-winged Sparrow *(Aimophila carpalis)* to a desert environment. Ph.D. dissertation. Univ. Ariz. 58 pp.
Rainfall is the breeding stimulus.

Phillips, A. R.
1939. The faunal areas of Arizona, based on bird distribution. M.S. thesis. Univ. Ariz. 62 pp.

———
1946. The birds of Arizona. Ph.D. dissertation. Cornell Univ. 498 pp.

Poore, J. T.
1969. The effects of water deprivation on the hypothalamic-hypophysial neurosecretory system of the Black-throated Sparrow, *Amphispiza bilineata.* M.S. thesis. Univ. Ariz. 19 pp.
Birds from Nogales, Arizona, were used.

Rea, A. M.
1969. The interbreeding of two subspecies of Boat-tailed Grackle *Cassidix mexicanus nelsoni* and *Cassidix mexicanus monsoni* in secondary contact in central Arizona. M.S. thesis. Ariz. State Univ. 131 pp.

Ricklefs, R. E.
1967. The significance of growth patterns of birds. Ph.D. dissertation. Univ. Pa. 118 pp.
Includes data on Arizona birds.

Robinson, M.D.
1968. Summer aspect of a high coniferous forest in the Chiricahua Mountains, Arizona. M.S. thesis. Univ. Ariz. 55 pp.
The elevational distribution of 23 species of birds is given.

Senteney, P.
1957. Factors affecting the nesting of Gambel Quail in southern Arizona. M.S. thesis. Univ. Ariz. 42 pp.
Near Tucson.

Sileo, L., Jr.
1970. The incidence and virulence of *Trichomonas gallinae* (Rivolta) in Mourning Dove (*Zenaidura macroura* Linnaeus) populations in southern Arizona. Ph.D. dissertation. Univ. Ariz. 34 pp.

Smith, E. L.
1967. Behavioral adaptations related to water retention in the Black-tailed Gnatcatcher *(Polioptila melanura).* M.S. thesis. Univ. Ariz. 43 pp.
Activity decreases during the hottest part of the day.

Soule, O. H.
1964. The saguaro tree-hole microenvironment in southern Arizona. II. Summer. M.S. thesis. Univ. Ariz. 75 pp.
Brief notes on birds.

Stair, J. L.
1970. Chronology of the nesting season of White-winged Doves *Zenaida asiatica mearnsi* (Ridgway) in Arizona. M.S. thesis. Univ. Ariz. 69 pp.

Steenbergh, W. F.
1967. Critical factors during the first years of life of the saguaro *(Cereus giganteus)* at Saguaro National Monument. M.S. thesis. Univ. Ariz. 51 pp.
Six species of birds eat fruit of the plant.

Straus, M. A.
1966. Incidence of *Trichomonas gallinae* in Mourning Dove *Zenaidura macroura* populations in Arizona. M.S. thesis. Univ. Ariz. 44 pp.
The average incidence was 20 per cent.

Taylor, W. K.
1967. Breeding biology and ecology of the Verdin *Auriparus flaviceps* (Sundevall). Ph.D. dissertation. Ariz. State Univ. 228 pp.

Tramontano, J. P.
1964. Comparative studies of the Rock Wren and the Canyon Wren. M.S. thesis. Univ. Ariz. 59 pp.
In the Santa Catalina Mountains, Arizona.

Truett, J. C.
1966. Movements of immature Mourning Doves *Zenaidura macroura marginella* in southern Arizona. M.S. thesis. Univ. Ariz. 60 pp.
The older immatures began the southward migration in July.

Van Devender, T. R.
1969. The effect of water deprivation at 32.3° C on the neurosecretory content of the *pars nervosa* of the White-crowned Sparrow, *Zonotrichia leucophrys gambelii.* M.S. thesis. Univ. Ariz. 18 pp.

Viers, C. E., Jr.
1970. The relationship of calling behavior of White-winged Doves to population and production in southern Arizona. Ph.D. dissertation. Univ. Ariz. 47 pp.

Westcott, P. W.
1962. The Scrub Jay in Arizona: behavior and interactions with other jays. M.S. thesis. Univ. Ariz. 33 pp.
Investigation chiefly at Oracle, Pinal County, Arizona.

Wolford, M. J.
1969. Vocal repertoire of the Cactus Wren *(Campylorhynchus brunneicapillus).* M.S. thesis. Univ. Ariz. 46 pp.
Includes sonograms.

9. Miscellaneous

Included here are reviews of books and papers relating to Arizona ornithology, obituaries of Arizona ornithologists, and biographies of ornithologists who worked in Arizona.

Anderson, A. H.
1965. [Review of] C. H. Lowe's The Vertebrates of Arizona. Auk 83: 116–117.

[Anon.]
1867. [Review of] Prodrome of a work on the ornithology of Arizona territory. Elliott Coues. Amer. Nat. 1: 209–210.

Archer, L., Mrs.
1960. "Yours, Jerry Stillwell." Aud. Mag. 62: 76–79, 82.
Brief mention of his recording of bird songs in Arizona.

[Austin, O. L., Jr., ed.]
1970. [Location and availability of the Bailey-Law natural history collection.] [In] Notes and News, Auk 87: 214.
At Rockbridge Alum Springs Biological Laboratory, Goshen, Virginia.

[Barlow, C., ed.]
1902. [Reports that T. E. Slevin is on a short collecting trip in Arizona.] [In] General News Notes, Condor 4: 25.

———
1902. [Reports that O. W. Howard and H. S. Swarth depart for a five months' collecting trip in Arizona.] [In] General News Notes, Condor 4: 49.

———
1902. [Reports that T. E. Slevin has been at Tucson, Arizona, for several months.] [In] General News Notes, Condor 4: 72.

———
1902. [Reports that H. S. Swarth and O. W. Howard return from a collecting trip into the mountains of Arizona.] [In] General News Notes, Condor 4: 123.

Bent, A. C.
1930. [Francis Cottle Willard—Obituary.] Auk 47: 455–456.

———
1937. In Memorium: Frederick Hedge Kennard, 1865–1937. Auk 54: 341–348.
Brief account of his visit to Arizona.

Bergstrom, E. A.
1964. [Review of] A Check-list of the Birds of Arizona. Gale Monson and Allan R. Phillips. Bird-Banding 35: 285.

Brodkorb, P.
1965. [Review of] The birds of Arizona. Allan Phillips, Joe Marshall, and Gale Monson. Bird-Banding 36: 286–287.

Brooks, M.
1938. Allan Brooks—a biography. Condor 40: 12–17.

R. C. [= Carlson, R., ed.]
1942. Birds of the desert. A review. [of Birds of the desert, by Gusse Thomas Smith.] Ariz. Highways 18 (1): 38–40.
Includes reproduction of sketches of birds from the book.

Clabaugh, E. D.
1930. Methods of trapping birds. Condor 32: 53–57.
Gila Woodpecker trapped in Arizona.

Clark, E.
1964. Arizona birds get the first full treatment. Ariz. Days and Ways Mag. Nov. 29: 44–45.
A review of Phillips, Marshall, and Monson, Birds of Arizona.

Colton, H. S.
1932. Samuel Washington Woodhouse. Mus. Northern Arizona, Mus. Notes 5 (1): 1–4.
A brief account of his Arizona travels.

———
1933. [Report on ornithological research, p. 39 in] 1932 at the Museum, Mus. Northern Ariz. Mus. Notes 5 (8): 37–40.

———
1934. [Report on ornithological research, p. 40 in] 1933 at the Museum, Mus. Northern Ariz. Mus. Notes 6 (8): 39–42.

———
1949. Hopi kachina dolls. Univ. New Mex. Press, Albuquerque, 144 pp. 1959. Rev. ed. 150 pp.
Brief notes on the feathers used.

Crockett, H. L. and R. Crockett
1964. A century of birding in Arizona. Ariz. Highways 40 (8): 34–39.
A brief historical account.

Cruickshank, A. D.
1966. Roadrunner in Tucson, Arizona. Cover photograph in Aud. Field Notes 20 (2).

Daniel, T., Mrs.
1965. Cactus Wren. Aud. Mag. 67 (1): cover photograph. Photographed near Tucson, Arizona.

Danson, E. B.
1965. [Mr. Harry King volunteers his services to the Museum in ornithology and is collecting birds.] [In] Annual Report Museum of Northern Arizona. Plateau, Suppl. 37 (4): 20.

———
1968. 40th Annual Report of the Museum of Northern Arizona and Research Center. Flagstaff, Arizona: 1–38.
Pp. 15–17, a brief account of research in biology.

———
1969. 41st Annual Report of the Museum of Northern Arizona and Research Center. Flagstaff, Arizona: 1–34.
Pp. 4–6, a brief account of research in biology.

Dick, H. W. and A. H. Schroeder
1968. Lyndon Lane Hargrave. A brief biography. [In] A. H. Schroeder. Collected papers in honor of Lyndon Lane Hargrave. Papers Archeol. Soc. New Mex. 1: 1–8.
Includes bibliography of his papers.

Dixon, J. S.
1942. Wildlife portfolio of the western National Parks. U.S. Dept. Int. Nat. Park Serv.: 1–121.
Good photograph of Long-crested Jay at the Grand Canyon.

[Edwards, H. H., ed.]
1928. Items from the Lawson Station: Oracle, Arizona. News from the Bird-Banders 3: 31.
A canary hatches and raises three House Finches.

Eisenmann, E.
1965. [Review of] The Birds of Arizona. Allan Phillips, Joe Marshall, and Gale Monson. Nat. Hist. 74 (6): 6.

E. [isenmann], E.
1966. [Review of] Phillips, A. R. 1964. Notas sobre aves mexicanas, III. Rev. Soc. Mexicana Hist. Nat. 25: 217–242. Auk 83: 343–344.

Ellis, E. H.
1929. [Dr. Loye Miller reports on his visit to Arizona.] [In] Los Angeles Chapter Meetings, [in] News from the Bird-Banders 4: 7.

[Emerson, W. O.]
1903. [Obituary of] Thomas E. Slevin. Condor 5: 57–58.
Brief mention of his stay in Arizona.

Emory, W. H.
1848. Notes on a military reconnaissance from Fort Leavenworth, in Missouri, to San Diego, in California, including parts of the Arkansas, Del Norte, and Gila rivers. 30th Congress, 1st session, Senate, Executive no. 7, Washington: 416 pp.
Itinerary of trip across Arizona.

[Fisher, W. K., ed.]
1903. [Arizona trips by W. W. Price.] [In] General News Notes, Condor 5: 25.

———
1903. [Obituary of] George H. Ready. Condor 5: 82.
Brief mention of his stay in Arizona.

Forrest, E. R.
1912. Photographing wild birds in southern Arizona. Ool. 29: 257–261.
At Oracle, Pinal County, in 1903.

Goldman, E. A.
1935. Edward William Nelson—Naturalist. Auk 52: 135–148.
Brief mention of his stay in Arizona.

Goodwin, J.
1965. [Review of] The Birds of Arizona. Allan Phillips, Joe Marshall and Gale Monson. Aud. Mag. 67: 128–129.

Greene, E. R.
1966. A lifetime with the birds. An ornithological logbook. Edwards Brothers, Inc. Ann Arbor, Michigan. 404 pp.
Includes brief comments on various birds observed in Arizona.

Grinnell, J.
1912. [Review of] The Birds of North and Middle America. Part V. Robert Ridgway. Condor 14: 110.

Guarino, J. L. and P. P. Woronecki
1967. Color marking to trace Blackbird and Starling movements. Western Bird Bander 42: 26–27.
Describes markings of birds banded in Arizona.

Hargrave, L. L.
1935. Ornithology, p. 34 [in] H. S. Colton, The Museum of Northern Arizona in 1934. Mus. Notes, Mus. Northern Ariz. 7 (8): 29–36.
Summary of work.

———
1936. Ornithology, p. 43 [in] H. S. Colton, The Museum of Northern Arizona in 1935. Mus. Notes, Mus. Northern Ariz. 8 (8): 39–46.

———
1936. Why birds are banded. Mus. Northern Ariz. Mus. Notes 9 (3): 13–16.
Banding at the Museum of Northern Arizona.

———
1937. Ornithology, p. 45 [in] H. S. Colton, The Museum of Northern Arizona in 1936. Mus. Notes, Mus. of Northern Ariz. 9 (8): 43–46.
A summary of work.

———
1938. A plea for more careful preservation of all biological material from prehistoric sites. Southwestern Lore 4: 47–51.

Hartrauft, C. H.
1908. An Arizona hunt. Ool. 25 (6): 85–87.
A running account of egg-collecting near Phoenix.

Heatwole, T.
1959. The birds know her address. Ariz. Days and Ways Mag. Sept. 27: 5–7.
An account of Mrs. J. T. Birchett's bird banding and her bird hospital.

Henshaw, H. W.
1920. In memoriam: William Brewster. Born July 5, 1851—Died July 11, 1919. Auk 37: 1–23.
Arizona field work by his collectors on page 21.

———
1920. Autobiographical notes. Condor 22: 3–10.

Holland, H. M.
1930. [Obituary of] Frances Cottle Willard. Ool. 47 (3): 33–34.
Mentions his Arizona activities. Santa Barbara Museum has his egg collection.

Horswell, J.
1959. [Field trips in Phoenix, Arizona area.] Ariz. Days and Ways Mag. Jan. 25: 19.

[Howard, O. W.]
1902. [Exhibition of Howard's Arizona collection.] [In] Official Minutes, Southern Division, Cooper Ornithological Club. Condor 4: 124.

Huey, L. M.
1938. Frank Stephens, pioneer. Condor 40: 101–110.

Hume, E. E.
1942. Ornithologists of the United States Army Medical Corps. Johns Hopkins Press, Baltimore. 583 pp.
Contains considerable material relating to Arizona in the biographies.

Johnsen, T. N., Jr.
1959. An epiphytic prickly-pear cactus. Ecol. 40: 324.
A seed in bird droppings is supposed to have started the plant in the crevice between the saguaro branches. Near Tucson.

Kiessling, V. R.
1938. The Kiessling station at Phoenix, Arizona. News from the Bird-Banders 13: 39–40.
Notes on bird banding activities.

King, D. S. and N. W. Dodge
1939. Banding at the Southwestern Monuments. News from the Bird-Banders 14: 13.
Brief summary of objectives.

Krutch, J. W.
1965. Living in the desert. Ariz. Highways 41 (3): 2–7.
Interesting comments on Roadrunners.

Laing, H. M.
1947. Allan Brooks, 1869–1946. Auk 64: 430–444.

Lanyon, W. E.
1957. [Review of] Birds of pine-oak woodland in southern Arizona and adjacent Mexico. J. T. Marshall, Jr. Wilson Bull. 69: 374–375.
Critical comments.

Law, Mr. and Mrs. J. S.
1928. A glimpse of Arizona. News from the Bird-Banders 3: 17–18.
An account of their visit to southern Arizona.

Linsdale, J. M.
1936. Harry Schelwald Swarth. Condor 38: 155–168.
Obituary with complete bibliography.

Manville, R. H.
1964. [Review of] The Vertebrates of Arizona. C. H. Lowe. Science 145(3629): 258.

Marshall, J.
1964. [Review of] The Song Sparrows of the Mexican Plateau. R. W. Dickerman. Auk 81: 448–451.
Includes a key to the subspecies.

McKee, E. D.
1939. Grand Canyon bird banding station. News from the Bird-Banders 14: 39.

Mearns, E. A.
1896. Ornithological vocabulary of the Moki Indians. Amer. Anthro. 9: 391–403.
The list contains 231 names.

———
1907. Mammals of the Mexican boundary of the United States. U.S. Nat. Mus. Bull. 56 (1): 1–530.
Included here only because of its list of collecting stations and itinerary.

M[iller], A. H.
1936. Frank Hands [Obituary]. Condor 38: 221–222.

[Miller, A. H.]
1948. [Note on Allan Brooks' colored frontispiece, opposite page 5, of the Zone-tailed Hawk in the Santa Rita Mountains, Arizona.] [In] Notes and News, Condor 50: 47.
Weight, stomach contents, and color of soft parts are given.

Miller, L.
1936. "Ft. Lowell, Arizona." Condor 38: 215.
Historical; photograph of the Post Trader's building in 1935.

———
1950. Lifelong boyhood. Univ. Calif. Press, Berkeley. 226 pp.
Pp. 55–72, Miller's Arizona collecting trip in 1894.

Monson, G.
1952. [Review of] Arizona and its bird life. H. Brandt. Wilson Bull. 64: 172–173.

———
1960. [Review of] The natural history of the southwest. Edited by William A. Burns. Aud. Mag. 62: 246.

———
1965. [Review of] The lives of desert animals in Joshua Tree National Monument. By Alden H. Miller and Robert C. Stebbins. Wilson Bull. 77: 417–418.
Includes a brief comment on Arizona birds.

R. E. M. [= Moreau, R. E.]
1965. [Review of] The birds of Arizona. Allan Phillips, Joe Marshall, and Gale Monson. Ibis 107: 264.

N[elson], E. W.
1913. [Herbert Brown: Obituary.] Auk 30: 472.

Nelson, E. W.
1913. [Herbert Brown: Obituary.] Condor 15: 186–187.

———
1932. Henry Wetherbee Henshaw—naturalist. Auk 49: 399–427.

[Palmer, T. S.]
1918. [Reports that H. S. Swarth visited southern Arizona in 1917 to study birds on the Apache Trail.] [In] Notes and News, Auk 35: 107.

———
1918. [Reports that E. A. Goldman collected in northern Arizona in 1917.] [In] Notes and News, Auk 35: 108.

———
1925. [Obituary of] Col. Harry Coupland Benson, U.S.A. Auk 42: 619–620.
Brief mention of Benson's ornithological work at Fort Huachuca, which began in 1884.

Palmer, T. S.
1950. [Obituary of] Frederick Monroe Dille. Auk 67: 548.
Brief mention of his stay in Nogales from 1938 to 1950.

Parkes, K. C.
1963. [Review of] Check-list of birds of the world. E. Mayr and J. C. Greenway, Jr. Wilson Bull. 75: 100–103.
Errors pertaining to Arizona are listed.

———
1966. [Review of] The birds of Arizona. Allan Phillips, Joe Marshall, and Gale Monson. Auk 83: 484–487.

Paynter, R. A., Jr.
1952. "Arizona! Magic is thy melodious name!" [Review of] Arizona and its bird life. Herbert Brandt. Ecol. 33: 313–314.

Peterson, W.
1954. Hunting Canadian Geese with a camera. Ariz. Highways 30 (2): 28–33.
At the Colorado River.

Phillips, A. R.
1940. Correction. Auk 57: 258.
Acknowledgments added to the paper in Auk 57: 117, 1940.

———
1940. Edgar Alexander Mearns (1856–1916), pioneer northern Arizona naturalist. Plateau 13: 1–5.

———
1946. [Review of] A distributional survey of the birds of Sonora, Mexico. A. J. van Rossem. Wilson Bull. 58: 56–57.
A critical review, with notes on Arizona species.

———
1950. [Charles Taylor Vorhies—Obituary.] Auk 67: 141.

———
1950. Charles Taylor Vorhies: 1879–1949. Science 112 (2909): 352–353.

———
1952. [Review of] Speciation and ecologic distribution in American Jays of the genus *Aphelocoma.* F. A. Pitelka. Wilson Bull. 64: 170–172.
Critical review, with discussion of Arizona records.

———
1953. Biology: the first twenty-five years. Plateau 26: 30–37.
Summary and bibliography of work at the Museum of Northern Arizona.

———
1966. [Review of] Biosystematics of sibling species of Flycatchers in the *Empidonax hammondii-oberholseri-wrightii* complex. Ned K. Johnson. Auk 83: 321–326.
Discusses Arizona specimens.

———, M. A. Howe, and W. E. Lanyon
1966. Identification of the Flycatchers of eastern North America, with special emphasis on the genus *Empidonax*. Bird-Banding 37: 153–171.
The useful keys include some species found in Arizona.

[Radke, E. L., ed.]
1967. Color-marking authorizations. Western Bird Bander 42: 49–52.
Includes Arizona bird markings.

Richmond, C. W.
1918. In Memoriam: Edgar Alexander Mearns. Born September 11, 1856—Died November 1, 1916. Auk 35: 1–18.
Brief mention of his work in Arizona, which began in 1884.

Robertson, H.
1903. [H. S. Swarth exhibits Arizona bird skins.] [In] minutes of club meetings, southern division. Condor 5: 27.

Robertson, J. McB.
1926. [Stephen C. Bruner bands birds at Fresnal Ranch.] News from the Bird-Banders 1 (3): 9.
Brief report of his activities 51 miles southwest of Tucson.

[Sargent, G., ed.]
1936. [Mr. Chambers reports on his trip through Arizona.] [In] Digest of Minutes of the Los Angeles Chapter of the W.B.B.A. News from the Bird-Banders 11 (1): 8.
He observed 500 Swainson Hawks and 6 Harris Hawks.

———
1936. Banding at the Southwestern Monuments. News from the Bird-Banders 11 (3): 27–29.
Includes a list of banders.

[Sargent, G., ed.]
1936. Banders carrying on special research. News from the Bird-Banders 11 (3): 33.
Includes a summary of the work of L. L. Hargrave at Flagstaff.

———
1937. [Victor R. Kiessling banding birds at Tempe, Arizona.] [In] News from Active Banders. News from the Bird-Banders 12 (1): 6.

———
1937. [Allan R. Phillips banding birds at Tucson, Arizona.] [In] News from Active Banders. News from the Bird-Banders 12 (1): 6.

———
1937. [Mrs. Partin reports on her trip to the Grand Canyon.] [In] Digest of Minutes of the Los Angeles Chapter of the W.B.B.A. News from the Bird-Banders 12 (4): 45–47.

Steele, E.
1966. Arizona's mystery bird. Aud. Mag. 68: 167–171.
Trogons in the Santa Rita Mountains.

[Steele, P. H., ed.]
1947. [A visit to Carlos Stannard of Phoenix and Mrs. J. T. Birchett of Tempe.] News from the Bird-Banders 22(2): 30.
A brief account of their activities.

Stephens, F.
1918. Frank Stephens—an autobiography. Condor 20: 164–166.

———
1936. [Obituary of] Charles E. H. Aiken. Condor 38: 91–92.
Brief mention of his collecting trip in Arizona.

[Stephens, T. C., ed.]
1936. [Review of] [Banding of Gambel Sparrows at Casa Grande Monument] Wilson Bull. 48: 326.

Stone, W.
1917. Grinnell on the Evening Grosbeak. Auk 34: 225.
A review of J. Grinnell's The subspecies of *Hesperiphona vespertina*. Condor 19: 17–22, 1917.

[Stone, W., ed.]
1917. [Reports that H. S. Swarth is now (1917) studying the breeding birds along the Apache Trail in southern Arizona.] Auk 34: 380.

———
1919. [Reports that he was engaged in field work in the Chiricahua Mountains from May 19 to August 1, 1919.] Auk 36: 635.

Swarth, H. S.
1903. [Swarth exhibits Arizona bird skins.] [In] Minutes of Cooper Ornithological Club Meetings, Southern Division. Condor 5: 27.

———
1909. [Review of] A. Wetmore's Notes on some northern Arizona birds. Condor 11: 73.
Swarth regards *Junco annectens* and *Catherpes mexicanus polioptilus* as dubious distinctions.

———
1911. [Review of] Notes on the birds of Pima County, Arizona. Stephen Sargent Visher. Condor 13: 37.
A critical review.

———
1911. [Review of] W. J. McGee's Notes on the Passenger Pigeon. Condor 13: 79.
Regards the report of its occurrence in Arizona as the result of a misidentification.

———
1911. [Review of] H. C. Oberholser's A revision of the forms of the Hairy Woodpecker (*Dryobates villosus* Linnaeus). Condor 13: 169–170.
Brief mention of Arizona races.

———
1911. [Review of] H. C. Oberholser's A revision of the forms of the Ladder-backed Woodpecker (*Dryobates scalaris* [Wagler]). Condor 13: 170.
Brief mention of Arizona races.

———
1925. [Review of] A monograph of the birds of prey, Part 1. H. K. Swann. Condor 27: 85–86.
Very critical of the failure to include recent discoveries.

———
1931. [Game bird fellowship at the University of Arizona.] [In] Editorial Notes and News, Condor 33: 40.

———
1931. [David M. Gorsuch in charge of Gambel Quail investigation in Arizona.] Condor 33: 40.

———
1931. [Review of] H. C. Oberholser's Notes on a collection of birds from Arizona and New Mexico. Condor 33: 81–82.
A critical review.

———
1935. [Review of] Life History of the Gambel Quail in Arizona. D. M. Gorsuch. Condor 37: 45–46.

Taylor, W. P.
1950. [Obituary of] Charles Taylor Vorhies 1879–1949. Jour. Wildl. Mgmt. 14: 96–97.

Tinker, F. A.
1953. Bill Carr. Outdoor professor. Aud. Mag. 55: 124–127, 139.
Mentions his activities in conservation in Arizona.

Tordoff, H. B.
1948. [Review of] The feeding and related behavior of Hummingbirds, with special reference to the Black-chin, *Archilochus alexandri* (Bourcier and Mulsant). F. Bené. Wilson Bull. 60: 253.

Townsend, C. W.
1925. Thrills of an eastern ornithologist in the west. Bird-Lore 27: 310–314.
Includes remarks on his visit to Arizona.

Tucson Audubon Society

1951 (?) The Roadrunner. Tucson Audubon Society, Tucson, Arizona. 15 pp.

A summary of public protests against the inclusion of the Roadrunner on the list of unprotected birds in Arizona.

Uren, L. S.

1965. Pellet gun kills Bonaparte, the rare Jay. Aud. Mag. 67 (3): 135.

Reports death of Magpie Jay [see Aud. Mag. 67 (2): 101–102.]

van Rossem, A. J.

1945. [Mention of recent trip to the Pajarita and Baboquivari Mountains, Arizona.] [In] Minutes of Cooper Club Meetings, Condor 47: 224.

Warren, E. R.

1936. Charles Edward Howard Aiken. Condor 38: 234–238.

Includes brief notes on his 1876 trip to Arizona.

Wehrle, L. P. and A. Phillips

1950. Bibliography of Dr. Chas. T. Vorhies. Jour. Wildl. Mgmt. 14: 98–99.

Includes many Arizona papers.

Wheeler, G. M.

1872. Preliminary report concerning explorations and surveys principally in Nevada and Arizona. 1871. Washington, Govt. Printing Office: 1–96.

Includes description of route in Arizona. No ornithological notes.

[Wolters, H. E.]

1965. [Review of] The Birds of Arizona. Allan Phillips, Joe Marshall, and Gale Monson. Jour. für Ornith. 106: 360–361.

Zimmerman, D. A.

1965. The Chiricahuas and Guadalupe Canyon. [In] The Bird Watcher's America. O. S. Pettingill, Jr., ed. McGraw-Hill, New York. pp. 312–324.

Personal experiences in these localities.

———

1966. [Review of] The Birds of Arizona. Allan Phillips, Joe Marshall, and Gale Monson. Wilson Bull. 78: 484–489.

10. Semi-popular

Illustrations constitute most of the items in this section. Many excellent photographs of Arizona birds in their natural habitats have been published in various popular magazines.

Allen, A. A.

1934. Blackbirds and Orioles. Nat. Geog. 66: 111–130.
Summaries of habits and distribution; also includes Shrikes, Vireos, Waxwings, and Phainopepla. Good paintings by Allan Brooks.

———

1935. The Tanagers and Finches. Nat. Geog. 67: 505–532.
Summaries of habits and distribution; paintings by Allan Brooks.

———

1937. The shore birds, Cranes, and Rails. Nat. Geog. 72: 183–222.
Summaries of habits and distribution; paintings by Allan Brooks.

———

1944. Touring for birds with microphone and color cameras. Nat. Geog. 85: 689–712.
Notes on birds of the Tucson area with a colored photograph of a Trogon.

———

1951. Stalking birds with color camera. Nat. Geog. Soc. Washington, D.C. 328 pp.
Good photographs of a number of Arizona species.

———

1954. Split seconds in the lives of birds. Nat. Geog. 105: 681–706.
Includes observations and colored photographs of birds of the Tucson region.

Bailey, F. M.

1939. Among the birds in the Grand Canyon country. U.S. Dept. Int. Nat. Park Serv. 211 pp.
A popular account with considerable new information.

Blackford, J. L.

1944. Desert dwellers. Nat. Hist. 53: 24–27.
Good photographs of several birds in the Hassayampa Plain, Arizona.

Blackford, J. L.
1944. Photographing Quail babies in the desert. Desert Mag. 7 (11): 17–18.
Photographs of Gambel's Quail eggs and newly hatched chicks "along Hassayampa River."

———

1946. They live in heat and drouth. Desert Mag. 9 (11): 5–10.
Photographs of desert birds, with some speculation.

———

1954. Wings in the desert hills. Desert Mag. 17 (2): 9–12.
A popular account with good photographs of Arizona birds in pinyon-juniper woodland.

———

1956. Wings in Saguaroland. Desert Mag. 19 (10): 24–25.
A popular account with good photographs of Sparrow Hawk and Great Horned Owl.

———

1956. Western wonderlands. A guide to bird habitats in the western United States. Vantage Press, New York. 120 pp.
Good photographs of Arizona birds, their nests and habitats.

Bleitz, D.
1955. Adventures with birds in Arizona. Ariz. Highways 31 (3): 6–31.
Twenty-three photographs.

Bohl, A. T.
1941. Desert neighbors. Ariz. Highways 17 (11): 36–39.
Good photographs.

———

1943. Wild Turkey in Arizona. Ariz. Highways 19 (4): 14–15, 43.
Notes on photography.

Bohl, W. E.
1948. Gambel Quail. Ariz. Highways 24 (4): 10–11.
Includes four photographs by H. L. and R. Crockett.

Booth, E. S.
1948. Birds of the west. Copyright by the author. 397 pp.
Popular. Keys, descriptions, and distribution, with brief notes on nests. 3rd ed. 1960. Outdoor Pictures, Escondido, Calif. 413 pp.

Bradt, G. McC.
1945. Birds of the desert spring. Desert Mag. 8 (7): 19–22.
Popular account with 12 good photographs of Arizona species.

Brandt, H.
1952. Arizona and its bird life. Ariz. Highways 28 (5): 16–29, with paintings and sketches from the author's book of the same title.
Autobiographical notes.

Brooks, A.
1934. Far-flying wildfowl and their foes. Nat. Geog. 66: 486–528.
Summaries of habits and distribution of Ducks, Geese, and Swans; includes some Arizona species. Good paintings by the author.

Bryan, K.
1925. The Papago Country, Arizona. U.S. Geol. Surv. Dept. Int. Water Supply Paper no. 499: 1–436.
Popular notes on birds; Gambel's Quail common.

Calder, W. A.
1968. There really is a Roadrunner. Nat. Hist. 77 (4): 50–55.
A popular account of his research on Arizona birds.

R. C. = [Carlson, R., ed.]
1951. Walter E. Bohl. Ariz. Highways 27 (3): 4–11.
Includes colored prints of his paintings of Arizona birds.

Crockett, H. L. and R. Crockett
1940. Arizona birds. Ariz. Highways 16 (3): 4–7, 25, 36.
Good photographs of birds and nests.

——— and R. Crockett
1941. Birds of the Grand Canyon. Ariz. Highways 17 (6): 16–19, 43–44.
Includes photographs.

——— and R. Crockett
1942. Arizona Hummingbirds. Ariz. Highways 18 (8): 14–17.
Good photographs.

——— and R. Crockett
1942. Cactus dwellers. Ariz. Highways 18 (11): 22–25.
Good photographs of birds and their nests.

——— and R. Crockett
1943. Gambel's Quail. Ariz. Highways 19 (2): 28–29.
Good photographs of birds and nests.

——— and R. Crockett
1943. More Arizona birds. Ariz. Highways 19 (5): 26–31, 37.

——— and R. Crockett
1945. Arizona birds. Ariz. Highways 21 (2): 18–27.
Thirty-one color photographs.

——— and R. Crockett
1946. Arizona Doves. Ariz. Highways 22 (2): 5–11.
Good photographs.

——— and R. Crockett
1948. Cactus Wren. Ariz. Highways 24 (3): 26–29.
Seven photographs, near Phoenix.

——— and R. Crockett
1948. Winter birds of the desert. Ariz. Highways 24 (10): 8–13.
Many good photographs.

——— and R. Crockett
1949. Arizona's Hummingbird parade. Ariz. Highways 25 (3): 4–11.
Good photographs.

——— and R. Crockett
1951. Desert nests. Ariz. Highways 27 (1): 8–11.
Good photographs.

Crockett, H. L. and R. Crockett
1958. Bird watching with a camera. Ariz. Highways 34 (1): 6–15.
Thirteen color plates; notes on photographing Arizona birds.

——— and R. Crockett
1964. Four seasons of birding in Arizona. Ariz. Highways 40 (8): 10–28, 30–33.
Contains many superb color photographs of birds.

Davis, E.
1961. Another look at Arizona birds. Ariz. Highways 37 (7): 16–25.
Thirty-three color photographs.

Dobie, J. F.
1958. "The Roadrunner in fact and folklore." Ariz. Highways 34 (5): 2–11.
A defense; very few Arizona items.

Dodge, N. N.
1938. Boarding house for birds. Desert Mag. 1 (4): 10–11, 28.
Popular account of winter feeding of birds at the Grand Canyon.

———
1964. Organ Pipe Cactus National Monument, Arizona. U.S. Dept. Int. Nat. Park Serv. Nat. Hist. Handbook Ser. no. 6: 1–91.
Good popular account with list of birds.

Edgerton, H. E., R. J. Niedrach and W. Van Riper
1951. Freezing the flight of Hummingbirds. Nat. Geog. 100: 245–261.
Includes color photographs taken in the Huachuca and Santa Rita mountains.

Finley, W. L.
1923. Hunting birds with a camera. Nat. Geog. 44: 161–201.
Includes photographs of Arizona birds.

Henshaw, H. W.
1915. American game birds. Nat. Geog. 28: 105–158.
General accounts of distribution with brief notes on several Arizona species.

———
1917. Friends of our forests. The Warblers of North America. Nat. Geog. 31: 297–321.
General accounts of distribution with brief notes on Arizona species.

Maslowski, K. H.
1947. "Red Tails." Ariz. Highways 23 (1): 34–35.
Notes on behavior, with three photographs of a nest in the Organ Pipe Cactus National Monument.

———
1947. The Elf Owl. Nat. Hist. 56: 166–167.
Contains photographs.

Museum of Northern Arizona Staff
1962. Oak Creek Canyon. Mus. Northern Ariz. 14 pp., paper, 3 photographs.
A popular account; Water Ouzel is listed as a resident along the creek.

Olin, I.
1948. Cactus campmates. Desert Mag. 11 (4): 27–28.
Popular account of the Gila Woodpecker, Curve-billed Thrasher, and Cactus Wren, with good photographs.

Pearson, T. G.
1932. The large wading birds. Nat. Geog. 62: 440–469.
Summaries of habits and distribution; includes Arizona species. Good color paintings by Allan Brooks.

———
1933. Crows, Magpies, and Jays. Nat. Geog. 63: 50–79.
Summaries of habits and distribution; includes Arizona species. Good paintings by Allan Brooks.

———
1933. Woodpeckers, friends of our forests. Nat. Geog. 63: 453–479.
Summaries of habits and distribution; includes Arizona species. Good paintings by Allan Brooks.

———
1934. Birds that cruise the coast and inland waters. Nat. Geog. 65: 299–328.
Summaries of habits and distribution of Pelicans, Grebes, Loons, and Cormorants, including some Arizona species. Good color paintings by Allan Brooks.

———
1936. Thrushes, Thrashers, and Swallows. Nat. Geog. 69: 522–546.
Includes Pipits and Horned Larks; paintings by Allan Brooks.

———
1939. Sparrows, Towhees, and Longspurs. Nat. Geog. 75: 353–376.
Summaries of habits and distribution; paintings by Allan Brooks and W. A. Weber.

Peterson, W.
1955. Family life of the Desert Quail. Nat. Hist. 64: 462–465.
Notes from vicinity of Phoenix; good photographs.

———
1956. Desert dandies. Ariz. Highways 32 (10): 28–35.
Excellent photographs of Gambel's Quail nests and nestlings.

———
1959. Vacationing with nature. Ariz. Days and Ways Mag. Jan. 25: 18.
Color photographs of Arizona birds.

———
1963. We've been stilted. Ariz. Days and Ways Mag. Aug. 25: 12–15.
An account of Black-necked Stilts nesting in the Salt River area.

———
1964. El Paisano. The story of the storied Roadrunner. Ariz. Highways 40 (8): 1–4, one colored plate.
On legends and facts.

Peterson, W.
1964. Water birds on the desert. Ariz. Highways 40 (8): 6–9.
A popular account of some Arizona birds.

Ratcliffe, W.
1963. I go hunting with my camera. Ariz. Highways 39 (10): 14–39.
Includes photographs of birds at Topock Swamp, Colorado River, Arizona.

Shantz, H. L.
1937. The saguaro forest. Nat. Geog. 71: 515–532.
Includes several photographs of nesting birds in the Saguaro National Monument.

Smith, G. T.
1941. Birds of the Arizona desert. Copyright by the author. 67 pp. A later edition is titled Birds of the southwestern desert, Doubleshoe Publishers, Scottsdale, 1956, 68 pp.
Contains some distributional data.

Sowls, L. K.
1963. Life in the desert. Aud. Nature Bull. Ser. 31 (2): 1–6.
A popular account with brief comments on Arizona birds.

Sprunt, A., Jr.
1954. Arizona's Cactus Wren. Ariz. Highways 30 (1): 28–33.
Eleven photographs by H. L. and R. Crockett.

Sutton, M.
1955. The valley of Doctor Mearns. Aud. Mag. 57 (1): 58–61, 80.
A popular account of some of the birds in the Verde Valley, Arizona.

Thornburg, R. and F. Thornburg
1951. Birds of Arizona's mountains. Ariz. Highways 27 (10): 6–9. 15 photographs.

Tippetts, K. B.
1932. Birds of the states. Amer. Nature Assoc. (?). 8 pp. not numbered.
Includes a colored picture, with brief account of the Cactus Wren, the state bird of Arizona.

Verplank, DeL.
1925. A sajuaro [sic] desert in Arizona. Nat. Hist. 25: 283–293.
Random notes on birds but no localities given.

Walker, L. W.
1945. Photoflashing western Owls. Nat. Geog. 87: 475–486.
Good color photographs of Elf Owls.

———
1950. Within the saguaro. Ariz. Highways 26 (4): 1–2.
Elf Owl observations.

———
1958. Arizona's window on wildlife. Nat. Geog. 113: 240–250.
Birds at a waterhole near Tucson.

Wetmore, A.
1932. Seeking the smallest feathered creatures. Nat. Geog. 62: 64–89.
Hummingbirds, Swifts, and Goatsuckers; includes Arizona species. Color paintings by Allan Brooks.

———

1933. The Eagle, king of birds, and his kin. Nat. Geog. 64: 43–95.
Summaries of habits and distribution; includes Arizona species; paintings by Allan Brooks.

———

1934. Winged denizens of woodland, stream, and marsh. Nat. Geog. 65: 577–596.
Summaries of habits and distribution of Titmice, Wrens, Nuthatches, Creepers, Kinglets, and Gnatcatchers, including Arizona species; paintings by Allan Brooks.

———

1935. Shadowy birds of the night. Nat. Geog. 67: 217–240.
Summaries of habits and distribution, including Arizona species; paintings by Allan Brooks.

———

1936. Parrots, Kingfishers, and Flycatchers. Nat. Geog. 69: 801–828.
Summaries of habits and distribution; also includes Cuckoos and Trogons; paintings by Allan Brooks.

———

1936. Game birds of prairie, forest, and tundra. Nat. Geog. 70: 461–500.
Summaries of habits and distribution, including Arizona species; paintings by Allan Brooks.

Zimmerman, D. A.
1962. Where to go. When to go. What to see. Aud. Mag. 64: 216–218.
Describes Guadalupe Canyon and its birds in Arizona and New Mexico.

———

1964. Choice birding in Arizona. Aud. Mag. 66: 191.
Brief description of the Chiricahua Mountains area and its unusual birds.